NB

Variations on a **Beehive**

ISBN: 978-1-912271-50-4

Published by Northern Bee Books © 2019

Northern Bee Books, Scout Bottom Farm
Mytholmroyd, Hebden Bridge, HX7 5JS (UK)

www.northernbeebooks.co.uk

Tel: 01422 882751

Book design by SiPat.co.uk

Variations
on a
Beehive

Compiled by

Patricia Nelson

Foreword

Richard Ball

The 'Devon Apicultural Research Group' or DARG, was formed 40 years ago. At the time there was concern that Devon BKA branch meetings were necessarily aimed at beginners and risked boring more experienced members. It was therefore considered that there was a need for an experienced beekeepers group, so a number of beekeepers got together and DARG was formed. Membership often extends into neighbouring counties.

The primary aim was, and still is, to collect, analyse and report on data, which may aid the advancement of beekeeping. Over the years many projects have been carried out by members covering such diverse issues including:

- A Devon swarm survey

- Drone drifting between colonies

- A survey of pollen collected by bees

- A survey of Drone Laying Queens and research into the reasons. This project was awarded the Vita Pinnacle Award.

Other activities include practical apiary demonstrations to show project techniques, project reviews and topical debates led by members or invited experts.

Further details can be found at *www.dargbees.org.uk.*

DARG aims to build on beekeepers observational strength and engage in research projects. Other activities are to further the apicultural education of its members and to disseminate

DARG's findings. To this end various leaflets were published in particular 'The Bee Way Code and 'Apiary Sites' both of which have been very popular with beginning beekeeping courses in Devon and elsewhere and the book 'A Case of Hives' edited by Len Heath, now out of print.

There is no doubt in my mind that new beekeepers do not give enough consideration as to the type of hive to suit their needs. Often the type used by the trainer is chosen and this can lead to all beekeepers in an Association using exactly the same type irrespective of their need. In my travels as a Bee Inspector I often asked beekeepers the type of hive they would use if they were to start over again, invariably it would be a different type!

In 2018 DARG were asked to update 'A Case of Hives' for republication. This edition is the result so, thank you to all the authors who have been selected for their knowledge of a particular hive. I encourage all new beekeepers to consider and if possible experience the various types before making a choice because in any beekeeping endeavour be it amateur or professional standardization should be the aim and not a mismatch of equipment. After all hives are expensive even though they have a comparatively high residual value.

Please enjoy this book and make the right choice.

Richard Ball, Chairman DARG, Sidmouth, May 2018

Contents

Experiments with hives

25. Developments in hive components

Introduction

Tricia Nelson

Much has changed for honey bees and beekeepers in the UK since Len Heath's "A Case of Hives" was published in 1985, not least the sheer number of beehive designs in use! As a result, Northern Bee Books asked the Devon Apicultural Research Group (DARG) to produce an updated version. This we have tried to do within the context of the many changes taking place. The following hive descriptions and photos have been provided by a variety of enthusiasts, without whose support this volume would not have been possible. We hope you will find their contributions both informative and interesting.

Some hives are designed to be used in a specific way and for a specific style of beekeeping whilst others can be adapted. A good example of an adaptation is the WBC as described, which has been adapted with an eke to use 14x12 frames. Some hives lend themselves to a variety of management techniques whilst others preclude them. Some hive designs have been revived and others such as the Cottager and Catenary appear to have fallen out of use.

Not all bees or beekeepers are the same. You will undoubtedly find differing opinions throughout the book. Always check references before relying on what you read and remember that certainty has a habit of being proved wrong! Choosing a hive is an individual decision and this is only to be expected.

There have been a significant number of other developments and inventions relating to beehives over recent years. These are set out in Chapter 25.

1. Insulation and hive materials
2. Floor designs as a response to pests
3. Foundation
4. Hive monitoring

Some of these innovations are as a result of improved opportunities, some based on research and some on personal experience.

Every design and innovation seems to have some advantages, but also disadvantages. It is for you the reader to form your own view and hopefully gain some useful insights into what both you and your bees need.

There are individual books and websites specific to some hive types and it is hoped this book will whet your appetite to read more about those you find interesting.

Tricia Nelson

Prologue

Tricia Nelson

Before we start it is important to provide a context within which to consider the different hives.

- European and American Foulbrood remain notifiable diseases in the UK.

- *Varroa* is endemic and better understood.

- Climate change is a live issue.

- Loss of habitat for both plants and animals including insects has been recognised.

- The unintended effects of certain pesticides and fungicides has reached the public consciousness.

- The declining numbers and species of insects.

- The Asian Hornet has now arrived in the UK.

- More materials are available with which to make hives.

- The age of electronic data has been upon us for some years.

- More information is commonly available and shared.

Choosing a hive for you and your bees

Most beekeepers choose the hive that their mentor or Association use, and only later start to wonder about whether this was the best choice for them, usually having already invested a considerable amount of time and money. Most of the contributors to this book are beekeepers who have decided to make a change.

So what is a hive? It is not the colony's nest. From the colony's point of view, it is simply a manmade cavity in which it can build a nest. Most European honey bee swarms prefer a cavity of about 40 litres (between 20 and 70 litres). This is the volume of a Langstroth deep (the brood box).

All beekeepers love bees, especially honey bees, and as we get to know them better, develop a profound respect for them, so our starting point with choosing a hive has got to be what the bees need.

Fortunately, we have the results of recent research into honey bees. This has provided more insights into colony life and honey bees' needs, including their preferred domicile when wild. Many people have wondered if the hives themselves could provide some possible solutions or benefits to colonies in their management of the ever-increasing threats against them? Others have wondered if hive designs have contributed to the problems that honey bees are currently facing.

The bees needs and preferences

European honey bees are basically tropical insects that have found a way to survive in the northern hemisphere. Part of their method is that they build their nests in cavities. Colonies vary in their behaviour and also the size of their colonies both as a result of strain and geography. It is now well recognised that locally adapted bees do best, so it is advisable to know your bees and your environment before choosing a hive.

In 1985 the bees' needs seemed fairly simple:

'*not made of material that will damage them, offers a reasonable defensive site against enemies and that there is sufficient space inside it for the colony*'.

All of which remains true. European honey bee colonies are incredibly adaptable and with adequate forage can live in almost any container, withstanding significant cold and heat. However, we now know that this comes at a cost and there are some key features which once understood by beekeepers suggest ways that hive structures might help a colony thrive.

1. **Thermoregulation**. Honey bees keep the brood nest at 34.5–36°C when there is brood. Any sustained deviation leads to damage to the sealed brood. This has now been thoroughly researched and how this temperature is maintained is better understood [10]. In short it takes a lot of energy and a lot of bees. Consequentially colonies start clustering at 18°C in order to preserve the heat they have created. This is an ambient temperature we perceive as quite warm, After all 'shirt-sleeve weather' is nearer 14°C or even cooler. When we then learn that a forager cannot warm its flight muscles to flying temperature of 27.5°C when its individual core temperature has dropped below 19°C, it becomes

even clearer how important it is to maintain heat in the nest. Did you know that if a colony can raise its brood nest temperature by 1°C it can help it combat disease? Is there a role for insulation?

2. **The importance of propolis.** There was a time when propolis was seen as a sticky nuisance. However, it is now known that it contains many chemicals that have significant antimicrobial properties as well as being used for waterproofing and sealing. Propolis collection seems to be highly beneficial to the health of the colony and is an intrinsic part of its immune system Bees expend a lot of effort collecting it and polishing the inner walls of their brood cells and cavity with it. Simply roughening the internal surfaces of the hive walls encourages the colony to collect more propolis. Would this be a good idea?

3. **Honey bees prefer small entrances.** Research has shown this has no detrimental effect on air flow, which they control themselves, and the entrance is easier to defend. Are our entrances too big? Does the position of the entrance matter?

4. **The shape of the cavity** is not as important as one might think. In the light of the following pages, this may come as a relief to many!

The beekeeper's needs and preferences

Physical and environmental considerations

Beekeepers have many individual requirements, some people are physically strong, some weak, some tall, others are short; some work as a pair, and some alone, some want to take their bees to different crops whereas others want a hive in the garden; some live in the country, some the city. Beehives can be kept on roof tops, hung in trees or placed in the corner of a field. Climatic and geographical conditions vary a lot across the UK. These are all important factors when choosing a hive design.

The reason for keeping bees

There was a time when most beekeepers primarily kept bees to make honey. This is no longer the most common reason for taking up beekeeping, and indeed many new beekeepers are somewhat confused when they discover there is honey that they could take and are not at all sure what to do with it, or even whether they should. Some want to study the bees, some want to offer safe haven to wild bees; for some it is a hobby, but for others a job and without them some of our crops would not get pollinated.

Degree of commitment

In addition to these factors a beekeeper must consider carefully how much time they are willing and able to commit to beekeeping, how many colonies they aspire to and whether building their hives is a possibility, an essential requirement, or completely out of the question. The final considerations, which are almost certainly the most important when making a final choice is how much money will it all cost and how long will beekeeping remain an active interest. Beekeeping is a capital heavy interest.

Tricia Nelson

References

- *The Nest of the Honey Bee* – T.D.Seeley and R.A.Morse, 1976

- *Honeybee Democracy* – T.D.Seeley, 2010

- *Darwinian Beekeeping* – T.D.Seeley, *American Bee Journal*, March 2011

- *Putting a Number on Natural* – D.Mitchell

- *Textured Hive Interiors Increase Honey Bee (Hymenoptera:Apidae) Propolis-hoarding Behaviour* – R.L.Hodges, K.S.Delaplane, Bery.J Bros, 2018

- *Ratios of Colony Mass to Thermal Conductance of Tree and Man-made Nest Enclosures of Apis mellifera: implications for survival, clustering, humidity regulation and Varroa Destructor* – D.Mitchell, *International Journal of Biometeorology* 60(5) pp629-638, 2015

- *High Humidity in the Honey Bee (Apis melliferaL) Brood Nest Limits Reproduction of the Parasitic Mite Varroa jacobsoni* – B.Kraus H.H.W.Velthuis, 1997

- *Honeybee Engineering Top Ventilation and Top Entrances* – D.Mitchell, *American Bee Journal*, August 2017

- *Winter Management* – William Hesbach, *Bee Culture*, 2016

- *The Honey Factory* – J.Tautz and D.Stein, 2018

- *The Lives of Bees* – T.D.Seeley, 2019

- *A Case of Hives* – Len Heath, 1985

- *Approved Document L Building Regulations 2016*, Ministry of Housing Communities and Local Government and product literature

1

The Development of the Beehive in the UK from the 19th Century

Will Messenger

I began my beekeeping when, as a young teenager, I went on holiday to stay with an aunt near Dartmoor. She would ask me to check whether the bees have any honey. I was directed to a veil and smoker in the barn and left to work things out for myself. The hive was a 'telescopic' WBC. In 1970 I took an optional beekeeping course as part of my teacher training. Now, the hives were Nationals, either the original part double-walled design or the still current Modified National. At one point I was working in London and commuting back to Somerset and the bees each weekend. I bought hive parts 'in the flat' from Lees

Fig. 1 – The engraving of the Renfrewshire Stewarton in B.B.J. for 21st July 1887 that inspired my interest in beekeeping history. My determination to one day keep bees in one of these hives was initially purely on aesthetic grounds. Note the double brood boxes, seven supers, and an eke.

of Uxbridge, made them up during weekday evenings and carried them on the train out of Paddington. American Foul Brood, self-diagnosed one January, wiped out my colonies, but long before that the seed of my interest in beekeeping history was sown with the chance purchase of a few volumes of the *British Bee Journal*.

After I retired from teaching I resumed beekeeping and quickly built up my stocks. I enjoyed a stint as a Seasonal Bee Inspector and, with my wife, Eve, ran a semi-commercial operation using mostly Modified National hives with home-reared bees. Many colonies were run on double brood boxes, some brood and a half, and a few deep National. Alongside this I returned to my latent enthusiasm for beekeeping history and am now secretary of the Beekeeping History Trust, having recently sold off much of the beekeeping business.

My brief for this chapter was to cover "all the hives that have not been written about" in the rest of the book. Just listing them would bust my word-count many times over, so I shall, instead, give an overview of hive development and focus on just a few types that illustrate the story.

Many myths about beekeeping in the past have become established fact. It is claimed that 'everyone' was obliged by the lord of the manor to keep bees, that all hives were straw skeps, and that the only way in which honey could be obtained was by killing the bees over the sulphur pit.

Certainly, many ordinary people lacked the skill, tools or resources to acquire timber and build hives, or the money to buy them; hive development was the province of the literate classes: the landed gentry and those that served them.

At least by the end of the 17th Century, the advantages of separating brood-rearing from honey storage had been recognised, as had the value of moveable frames, a system for excluding the queen from honey supers, and the provision of top entrances. Whether all these ideas were ever put together in one hive is uncertain, but the essentials of hive design that allowed modern beekeeping practice were there well before our period. When reviewing hive design from the past, it is important to remember that peoples' motivations for keeping bees were not necessarily our own. Most famously, Charles Darwin was only interested in gathering evidence in support of his 1859 work On the Origin of Species, and his observation hives were quadruple-glazed.

Fig.2 – Francois Huber's Nouvelles observations sur les abeilles was first published in 1792 with the first English (Scottish!) edition in 1806. This example of Huber's leaf hive, never intended as a hive for honey production, is held in the National Museum of Rural Life, Glasgow.

Plate 1.

Fig. 1.
Fig. 2. Fig. 3.
Fig. 4.
Fig. 6.
Fig. 5.

A

TREATISE

ON THE

Breeding and Management

OF

BEES,

TO THE GREATEST ADVANTAGE.

Interspersed with Important Observations,

ADAPTED TO GENERAL USE.

DEDUCED FROM A SERIES OF EXPERIMENTS DURING
THIRTY YEARS,

BY JOHN KEYS.

A NEW EDITION.

London:

PRINTED FOR LACKINGTON, ALLEN, AND CO.; F. C.
AND J. RIVINGTON; LONGMAN, AND CO.; C. LAW;
J. WALKER AND CO.; R. BALDWIN; D. AND R. CROSBY
AND CO.; SHERWOOD AND CO.; AND T. HAMILTON.

1814.

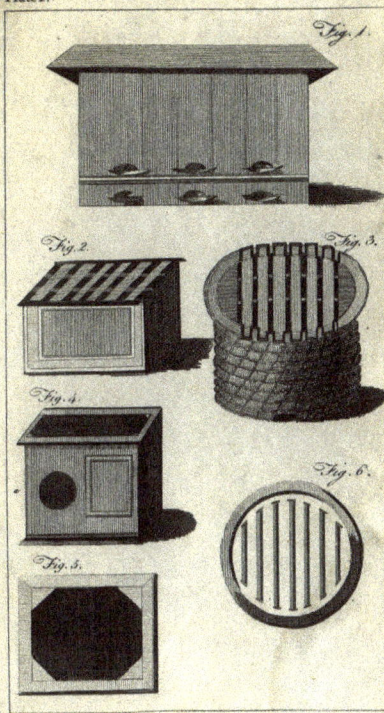

Fig. 3 – Title page and frontispiece of John Key's 1814 Treatise on the Breeding and Management of Bees. This shows that top bars, carefully positioned to respect the bee space, had been developed, and that many hives were intended to be kept inside bee houses or under bee shelters.

By 1800, stimulated by the development of the microscope, there was widespread interest in entomology and the science of the honey-bee, so hives that we would consider utterly impractical for honey production were popular for showing the work of the colony. The Reformation had started the decline in importance of beeswax and this was accelerated by the increasing availability, and cheapness, of mineral oils for lighting. Industrialisation and periods of economic depression generated a landless and low paid underclass and drove interest in beekeeping for honey-production as a means of advancing the lot of the rural poor. At the same time, technology introduced new materials and manufacturing methods and opened up business opportunities. Progress in working conditions gave ordinary people spare time and rising literacy and cheaper printing and publication costs brought information to the whole population.

John Keys, in his 1814 book, clearly understood bee space: "bars ... laid across the top of the hive, at half an inch distance from each other; the outermost to be one inch and a quarter wide, and the others one inch and a half". By the beginning of the C19th, humane beekeeping, i.e. recovering a honey crop without harming the bees, was understood, but the principle was most famously developed by Thomas Nutt in his 1832 *Humanity to Honey Bees*. Nutt introduced his collateral hive after about ten years development, claiming, along with W. Dyer in 1781, complete originality, although Rev. Stephen White had published *Collateral Bee-boxes* in 1756.

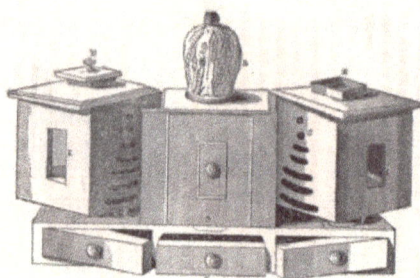

Above: Fig. 4 – Nutt's Collateral – National Museum of Rural Life, Glasgow.

Right: Fig. 5 – Nutt's Collateral. Plate taken from his book with "every compartment exposed to view".

Figure 5, taken from Nutt's book shows the three bee-boxes, with movement between them managed by 'dividing tins'. Thus the brood will normally occupy the central box and the honey the outer two. A crop could be taken by closing one of the outer boxes with the divider and allowing the bees to return to the centre. Nutt also advised that a swarm could be taken in one of the outer boxes and then united back to the main colony. Despite his opposition to 'storifying', Nutt added glass jars on top of each box to allow further expansion of the honeycomb.

The principle of storifying was well established by this time, having been an important aspect of the Stewarton system introduced by Kerr in 1819. Nutt devoted a whole chapter to 'Objections against Piling Boxes' and others advocated nadiring rather than supering; but with an arrangement of fixed top bars allowing sliders between to manage the movement of bees from box to box, the Stewarton allowed the queen to be confined to a brood chamber. John Keys, in his *Treatise on the Breeding and Management of Bees* (1814) made the key point that: "A good storifier that has not swarmed, or has had the swarm returned, will increase thirty pounds in seven days...: whereas a single hived stock in the same apiary and season, that has swarmed, will not increase above five pounds in the same time".

Fig. 6 – Two original Stewarton Hives from the IBRA collection

A full description of the Stewarton system is beyond the scope of this chapter. Despite the similarity to windows in many other hives, those in the Stewarton are not for observation but are integral to the management. Manipulation of the slides allows control of movement of workers (but not the queen) into the honey supers so that comb is built from the outside towards the centre; when bees can be seen occupying comb and obstructing the light, it is time to add another super. Thus development of the colony during the season can be entirely managed without opening the hive – something that is impossible in all modern hives.

Apart from the straw skep, this hive is probably the longest-lived of all. Stewartons were in regular use up to the 1970s, Rothamsted experimented with one around that time, and we manufactured and used an enlarged version during this century. Others continue this enterprise to the present.

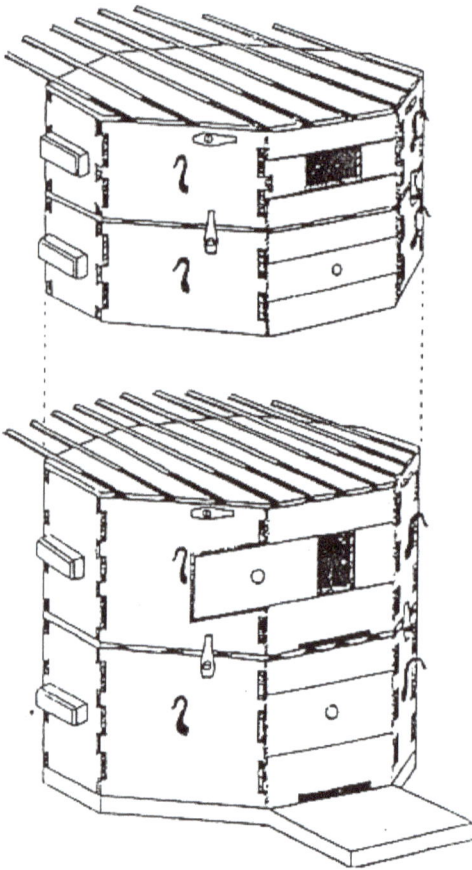

Above: Fig. 7 – The hive introduced by Kerr of Stewarton in 1819, drawn by 'TLK' for an article by James Struthers delivered in 1951. Note the different spacing of the top bars in the brood chambers and the supers.

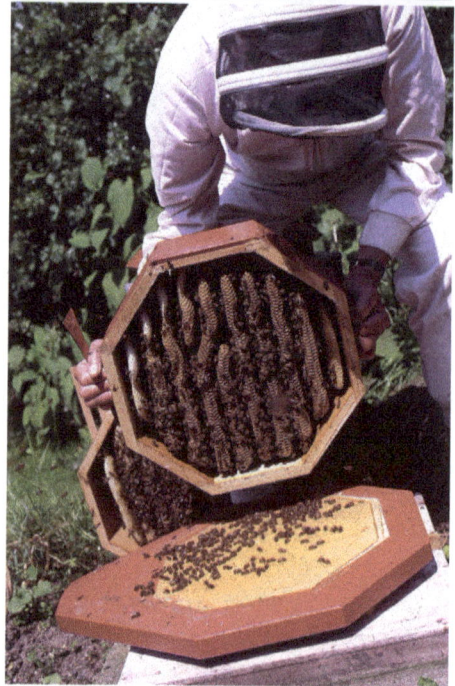

Above: Fig. 9 – One of our 'modern' Stewartons and it's owner enjoying experimental Victorian beekeeping.

Left: Fig. 8 – The colony we kept in a Stewarton for more than seven years – note the clear observance of bee space!

Beekeepers are taught that Langstroth 'discovered' the bee space and invented the modern moveable frame hive. In 1838 Major Munn patented an essentially similar hive, but in French, and in Paris, before Langstroth. The 1851 Great Exhibition was a good showcase for modern beekeeping and numerous box hives, i.e. with top bars, were available. Thomas White Woodbury is credited with introducing the bar and frame hive to Britain.

Fig. 10 – One of William Augustus Munn's hives, National Museum of Rural Life, Scotland. The illustration shows the moveable frame and the cover to allow any frame to be lifted out of the hive for examination.

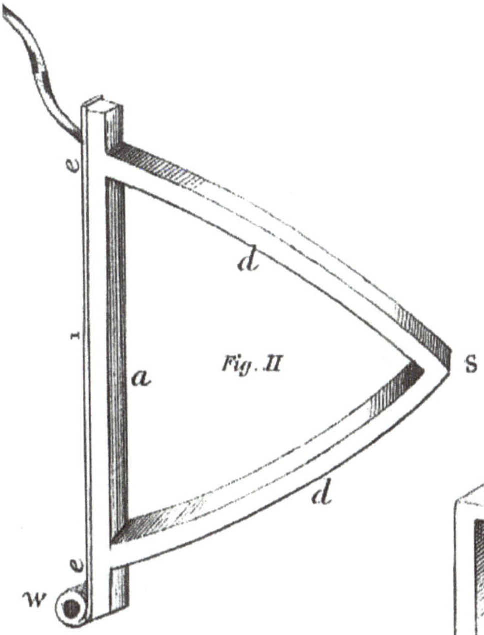

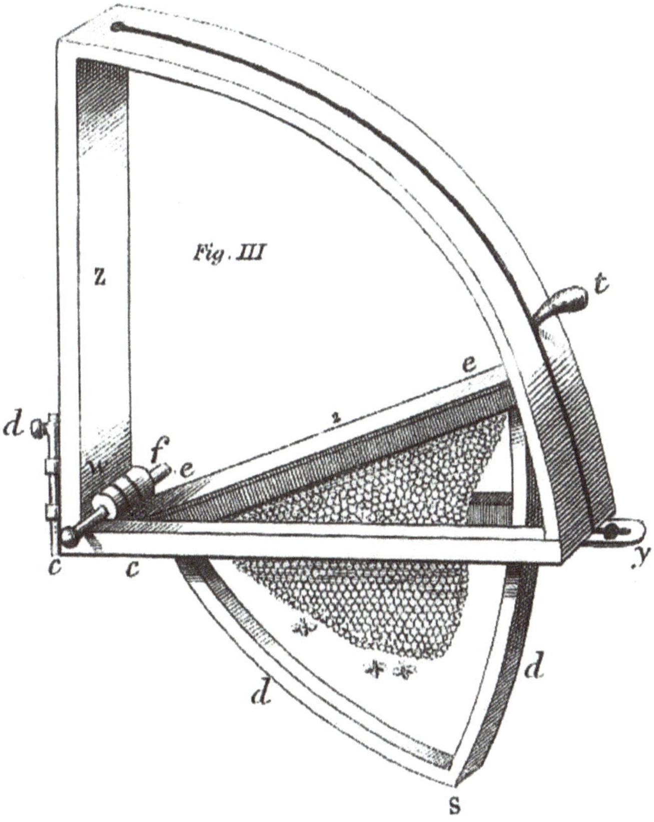

Fig. II

Fig. III

NEIGHBOUR'S IMPROVED SINGLE BOX HIVE.

We have introduced the "Single Box Hive" to suit the convenience of those who, though desirous of keep-

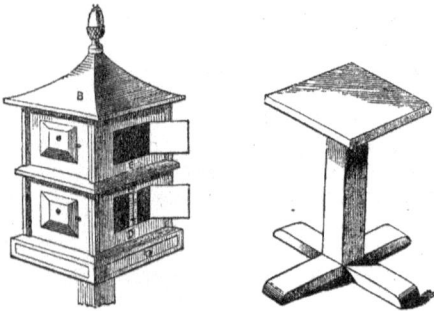

i ng bees on the improved principle, do not wish to incur the expense or devote the space which is necessary for Nutt's hive.

WOODBURY STRAW BAR AND FRAME HIVE.

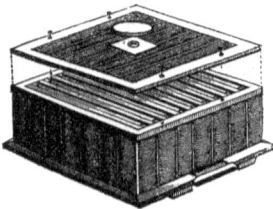

Since the introduction of the wood hive by Mr. Wood-bury, that gentleman has recommended, in the *Journal of Horticulture*, that the stock-hive be made of straw, of

Fig. 11 – A selection of mid-to-late C19th hives featured in Alfred Neighbour's The Apiary

THE LADIES' OBSERVATORY OR CRYSTAL BEE-HIVE.

The following engraving illustrates the construction of the Ladies' Observatory Hive. The stock-hive is cylindrical, with a flat top and a hole in the centre; the dimensions twelve and a half inches inside, eight and a half inches deep; the outer cover being raised, and made of stout glass, so as not easily to break. A support, composed of even wooden bars fixed on a pedestal from the floor-board, is very useful for the bees to cling to and attach their combs, instead of resting wholly against the glass.

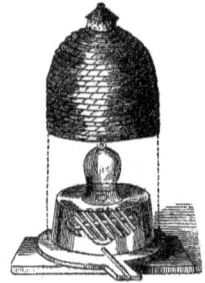

other boxes, being five inches deep. In many localities and seasons, the third box may not be required. Each box has two windows, one at the back and another at the side, a zinc shutter, sliding in a groove, excluding

Despite developments in wooden hives during the nineteenth century, straw was still an important material, due in part to its ready availability, but also for its insulating and 'bee-friendly' properties. Neighbour's Improved Cottage Hive, publicised in Alfred Neighbour's The Apiary, of 1865, is probably the pinnacle of design for this type. The surviving example owned by Sussex BKA has (probably sweet chestnut) hoops that are still perfectly circular after about 150 years.

Following launch of the B.B.J. in 1873 and the formation of the B.B.K.A. in 1874, numerous hives were introduced to beekeepers – the Abbott, the Cowan and Raynor hives among others. In 1892, at the B.B.K.A. Conversazione, George Wells, of Aylesford, Kent, introduced his 'New Method of Keeping Bees'. This involved keeping two colonies in one large brood chamber separated by a perforated divider but sharing a common storage space above a queen excluder. Numerous manufacturers produced their own version of the Wells hive

Fig. 12 – Neighbour's Improved Cottage Hive showing the inner hoop and queen excluders to prevent brood being reared in the glass supers.

Fig. 13 – Neighbour's Improved Cottage Hive – the outer cap complete with ventilator, the glass supers, and the maker's plate.

Fig. 14 – The surviving Wells Hive made by Charles Redshaw of South Wigston, Leicester.

including W.P.Meadows, Howard, A. W. Harrison, Thomas Bates Blow and Neighbour & Son. The secret of Mr. Wells' success in producing massive honey crops seems to have been to ensure that one of the queens was young, a point overlooked by many beekeepers. William Herrod-Hempsall was happy to testify that Mr Wells could, and did, successfully work his system, but added that he was the only man who could! Others denounced the hive as yielding "neither pleasure, nor profit". They feature in many photographs of apiaries of the time, but were reckoned to end up as chicken coops or dog kennels!

Fig. 15 – The Burgess Perfection in summer configuration

J. T. Burgess & Sons introduced their Perfection Hive which was truly telescopic, allowing bees to be wintered behind four walls – ideal for the warm/wet climate of the area.

HIVES WITH LARGER BROOD CHAMBERS.

THE EASY TO WORK HIVE

This Hive Gives a Larger Broodchamber

The broodchamber of this hive will hold 14 frames which allows for a breeding area of about 80,000 cells, or nearly half as much again as in the standard ten-frame broodchamber. This extra room has great advantages. It enables swarming to be controlled more easily, as the colony has room to expand without having a second broodchamber added. It is the hive we use in our own apiaries, and our usual practice is to leave sufficient frames to accommodate the winter cluster, add extra ones in the spring as required, and when swarming time comes to make nuclei and so check the swarming impulse. Most of the year the brood-chamber therefore has 10 or 12 frames only, and when handling the bees the dummy can be drawn to the back of the broodchamber, leaving more room in which to lift out the frames. The hive takes ordinary supers, and when these are in use the frames at the back of the broodchamber are kept covered with a folded quilt.

SPECIFICATION AND PRICES

The hive is sent out as shown, and is fitted with twelve standard frames, dummy, queen excluder, super of sections or shallow frames and quilts.

	No.	Made up	No.	In the flat
Hive as specified	190	48/-	190a	39/6
Do. with foundation in broodchamber ...	193	58/6	193a	44/9
Do. Empty, no frames or super	196	34/6	196a	28/6

If fitted with foundation throughout, the prices will be, with section super No. 199 : **60/2**, or with shallow frame super No. 194 : **64/2**. Painting 8/- per hive extra.   For prices of spare parts, see page 3.

IF USED AS A FOURTEEN FRAME HIVE THROUGHOUT

Supers to hold 14 frames will be required and we can supply these as follows :

	No.	Made up.	No.	In the flat
Empty 14 frame super with plinths, for deep frames	191	7/6	191a	6/-
Do. for shallow frames	192	6/-	192a	4/6

190

CROSS SECTION OF HIVE 190

Illustration of Burtt & Son EASY TO WORK hive – extract from Burtt's 1938 Catalogue (page 5)

Fig. 16 – The advertising for the Burtt's Easy to Work Hive

Burtt & Sons introduced their Extra-Deep, Easy to Work Hive. This could certainly live up to the claims made for it, but was inspired by wanting to make use of longer planks coming through the workshop.

Fig. 17 – A surviving Burtt's Easy-to-work hive – some are still in use in the Gloucester area.

And finally, a representative of the post-WW2 boom in hive design. There are a few aluminium hives to be found, dating from the timber shortage of the time, but I know of no surviving example of the Reynolds hive.

Fig. 18 – An idyllic representation of an apiary of Reynolds Hives – do any survive?

In conclusion, the Reynolds Hive picks up Thomas Nutt's disdain for tiering; hive design is subject to the ideas of the beekeeper conditioned by the technology, materials and thinking of the times. Beekeepers will always debate the pros and cons of different hives, but the bees will be bees - plus ça change, plus c'est la même chose!

© *Will Messenger, June 2019*

2

Bee-skeps

Chris Park

My name is Chris Park and I keep bees on the Oxfordshire/Wiltshire border. I keep them for pleasure, health & research. Apparently its unlucky to count, but I have around 20 something hives stocked with bees. I have kept bees for 10 years, housed within Skeps, Nationals, WBCs, Langstroths, Top bar hives, Log hives, Warré hives, Dartington hives and a Bee House.

A 'skep' is simply a basket. 'Bee-skeps' are baskets for bees.

They have survived in Britain and Ireland primarily as swarm collection vessels. In recent years, there has been a slow resurgence in the craft of skep-beekeeping. Commonly recognised as symbols depicted on the side of honey jars, bottles of mead, banks, pubs and various societies fostering community and co-operation. They have a certain charm about them.

A swarm of bees that is shaken, brushed or otherwise ushered, into a straw bee-skep is generally happier there. They have shelter from any breeze, protection from the sun or rain, and there are many places for them to get a foothold. They can easily spread out, diligently creating more opportunity to fan their Nasanov pheromone to call in returning scouts, and those left upon the bough or bike-shed.

Straws, being hollow, create good insulation, and the bee-skep's shape usually being rounded, suits the bees' natural cluster. The materials also breathe. It is interesting to note that two of the traditional materials that are most commonly split to make the lapping, that binds the straw, are both plants that benefit bees greatly: Willow and Bramble.

Form follows function. The archetypal bee-skep shape is a kind of bell-like, almost parabolic dome. It is a continuous spiral of coiled straw from top to bottom. Every 'rowl' of straw rests upon the next. Bee-skeps may also be fashioned from wicker and plastered with a composite material like cow-dung or daub. When permanently positioned, a bee-skep of any kind is traditionally sealed to its floor using a similar substance.

This one pictured at the top of the next page is akin to Charles Butler's (1571-1647) preferred form, being slightly egg-shape; tapered in at the base. He was a vicar in Hampshire, U.K. who wrote a famous and comprehensive book on skep beekeeping[1] in the early 17th century, commonly known as 'The Feminine Monarchie'. It is a fascinating read. Within it he calibrates the year astronomically, saying things like "At *Gemini* therefore set the doores wide open…", "In Leo though the swarming time be past…" and "To remove in *Virgo* is dangerous for robbing…". Some editions contain music and song that Butler composed for his bees. However, for most part, it is a 'how to' for the budding and literate beekeepers of his age.

It is perhaps due to his book that I often hear knowledgeable beekeepers say, on the subject of skep-beekeeping, "you can't do that anymore can you. You have to kill the bees to harvest

---

1 | *Butler,C. The Feminine Monarchie. Barnes. Oxford. 1609*

the honey!". Well I hope here to dispel that myth. It was indeed one way to harvest from a bee-skep, and it was Butler's preferred method, but not the only way. He did indeed write about the other methods, and more have come into practice since[2]. He obviously liked the no fuss procedure of placing the heaviest and lightest stocks over a pit, containing a lit sulphur packet, to suffocate the bees and the brood before he cut out the combs. This seems barbaric by today's standards, but he was writing at a time before human rights, let alone animal rights. We do indeed still cull bees today to thwart foul broods, and I'm sure it would be alarming to learn how many un-saveable or unwanted colonies are euthanised by pest control officers annually.

I imagine prolific beekeepers who kept many stocks in Butler's day (his own model apiary containing around 56 when full) viewed their bee-gardens a bit like a forester viewed a woodland or coppice. They knew if they harvested a bit here, and another bit there, it would all come springing back next season. I'm sure that somewhere in their minds was the notion that it meant a few less swarms to deal with next year!

Well, rest assured that one can successfully keep bees in skeps, whilst artfully ensuring happy and healthy stock, and with a good harvest to boot, without suffocation, and with minimal fuss.

The question one ought to answer first is perhaps, 'why?'.

---

2 | *For example: Isaac, J. The General Apiarian. 1803. Exeter*

I often say that "why one keeps bees often determines how one keeps bees", and therefore to a certain extent which hive one might choose.

I am also guilty of stating that "beehives are for beekeepers, and not for bees" and "bees don't mind what they are housed in if they have enough room and are well protected from the elements, pests and predators". It is the "spirit with which they are kept that is most important". To contradict myself though, I like to think that bees love a well-made bee-skep, housed within a good shelter, above most other hives.

Why do I keep bees in skeps? There are a multitude of excellent reasons. I have a background in ancient technology, primitive craft and experimental archaeology. Practicing skep-beekeeping preserves an ancient craft and updates it to modern pests and diseases. It is a part of our beekeeping heritage. It enables one to fashion a hive for bees with what is available from the hedgerows and meadows, being a sustainable and resilient style of low-tech beekeeping. It is a rewarding avenue of research. There is much for us to re-learn. It is educational and inspiring. It can be beautifully bee-centred. It can ensure low stressed and healthy bees.

Whatever one's reasons might be for stocking a bee-skep, it needs a shelter. The symbol of a busy skep, sitting in sunlight, surrounded by a verdant array of flora says many things to one's soul. It is however misleading, as far as practical skep-beekeeping is concerned. It would not last a year if unprotected. Probably the most well-known skep shelters are bee boles. Often found in groups, they are a collection of alcoves in a wall, which may be stone, brick and sometimes cob. They may have winter doors, and some are found in cellars as winter quarters.

Other simple shelters used were old creaming pans, pots and sacks and suchlike. Another common practice was an open sided shed or lean-to. Perhaps though the most charming and ancient skep shelter is known as a *hackle* or a *coppet*. This is a conical hat usually made of straw, and sometimes reed or willow. It sheds rainwater like thatch and buffers the colony from the cold wind in the winter and the hot summer sunshine.

The simplest way of harvesting honey from a bee-skep is cutting a piece of honeycomb out from the smaller side combs. This is sometimes referred to as 'castrating' the comb. A cottage beekeeper with only a handful of colonies would simply cut some more when the honey pot in the larder was empty. An 'Eke' (an extra section of about 5 coils of straw) may be added underneath for drawing more comb down into.

Another technique used is to 'drive the bees' from a skep that is full. This involves up turning the colony and pinning an empty skep to the top with driving irons: two iron staples and an iron pin, or hazel spars and a hazel skewer. The lower skep is then rhythmically drummed upon, a tiny bit of smoke can help at the beginning, and it is best performed a few metres away from the stand. The bee's instinct in this state of emergency is to walk orderly upwards, away from predators and damp, to a dark place. After a minute or so, a handful of bees have ascended and have begun to fan. After about 20–30 minutes all the bees have occupied the top skep, which is placed onto the original stand. They have the rest of the season to build themselves up ready for the winter. From the original skep, honeycomb, wax and brood can be harvested.

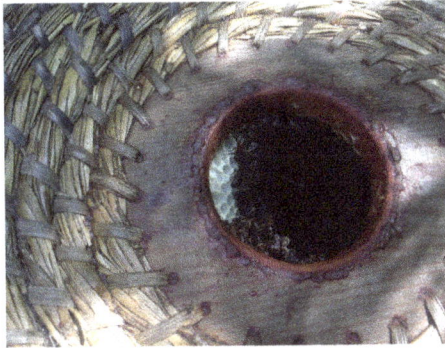

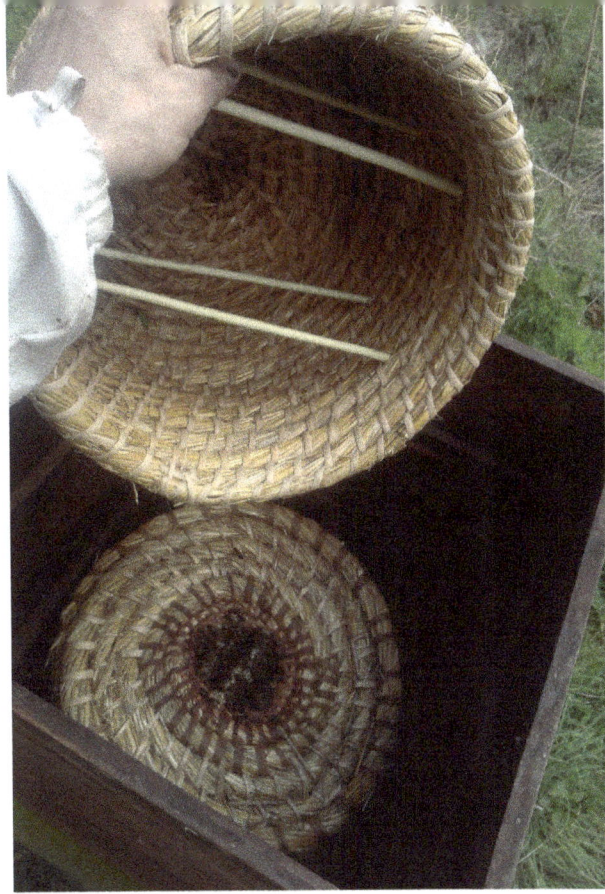

This procedure may also be done in the Autumn, and the bees may be united with another colony. Driving the bees has been a common 'show' practice within regional associations and at club meetings, exams, summer gatherings or end of season events. Again, this is making a slow come back, and another reason one might keep some bees in a skep.

Bee-skeps need 'sticking'. There are many ways to stick a hive. The most common is by driving sharpened hardwood spars through the side wall with a hammer and into the opposite wall. These act as brace comb and support the basket when the weight of the colony increases.

One skep-beekeeping system that A. Pettigrew (a.k.a. the bee-man's son) wrote comprehensively about, that will be familiar to conventional modern beekeeping, involves multiple layers of skeps, like a brood box and supers[3]. The bottom, usually flat topped, bee-skep goes through the winter, and has a hole in the top that is opened when a 'super' or 'cap' is placed on top. At the end of the season the top basket is removed and only honeycomb and wax is harvested. There is no need for a queen excluder, simply a three-inch diameter hole in the top of the brood skep for the bees to travel through and extend the comb. Only once have I seen some drone brood laid above the hole, and by the end of the season it had been replaced with honey stores.

How is skep beekeeping relevant and of value to contemporary beekeeping? It grew out of fashion when moveable frame hives became popular. Were some babies thrown out with the bathwater?

---

3 | Pettigrew, A. *The Handy Book of Bees: Being a Practical Treatise on Their Profitable Management*. Blackwood & Sons. 1870

There are some obvious disadvantages to the contemporary beekeeper:

1. The difficulty in inspecting brood comb.

2. An inability to use foundation or re-use drawn out supers. (Though one could put starter strips in, there is no need)

3. Swarm management. Difficulty in interpreting or preventing a swarm.

4. The honey harvest. There are no neat wooden frames to take hold of. Combs can be irregular. It can be a stickier process. Honey is drained and/or pressed instead of spun. (Though some apitherapists might say this is an advantage considering the honey's characteristics)

If one were only keeping bees in skeps, and harvesting from them, the amount of kit needed is minimal. No frames, foundation nor honey-spinner etcetera are required. The craft can be much simpler in that respect. I would say that the extra care involved in dealing with swarms and harvesting balances well the minimal fuss of the craft in general.

I must also add that I have never treated bees with chemicals of any kind. Thus, allowing them to develop their own strategies to combat pests and disease. Part and parcel with this practice is allowing them to swarm. This is usually the way with bee-skeps, ensuring early swarms, re-queening early and forever building new comb. Therefore, the siting of a skep apiary needs careful consideration.

I rarely suffer winter losses but do have summer losses. These occur through missing some swarms, and virgin queens not mating or returning to hive. Again, if a colony becomes queenless, there is opportunity to renew comb. I site lure hives around the farm that do catch swarms.

I keep many other styles of hive, from logs to Langstroths. They are all managed in a similar way: no treatments and free to swarm. The moveable frame hives can be inspected and can be considered as a kind of control.

Here are just some of the advantages found within skep beekeeping, and not all are exclusive to skeps:

1. A rounded form suits the shape of the bees' cluster. This eliminates cold corners and enables the vertical, oval shape of the brood to develop naturally.

2. They are breathable and well insulated.

3. The bees draw pure virgin comb. Reducing traces of chemicals, miticides, fungi or other waxes from recycled wax foundation that may be affecting the fertility of queens and drones.

4. They are their own planners and architects. Waxen combs are orientated the way the bees choose, to their best advantage, in response to their environment, for the management of ventilation and logistics. In my experience, they draw combs the cold way.

5. The bees decide upon their own cell size from the onset.

6. Minimal inspection means minimal stress.

  6.1. The colony's scent is disturbed less.

  6.2. The colony's temperature is maintained (consider *Varroa*).

  6.3. There is minimal risk of spreading disease (beekeepers are the biggest vector for bee disease).

  6.4. Less propolis is damaged. A propolis curtain can be built across the entrance, sometimes being 'closed up' for winter like a mouse guard with two small holes in.

  6.5 Bees are squashed less (consider Nosema).

  6.6 Bees are smoked less.

  6.7 Comb is broken less.

  6.8 The colony undergoes minimal exposure to light.

7. The honeybees swarm. This is a natural way of re-queening, an effective *Varroa* control, and one of the bees' own disease management strategies. This ensures new comb and renewed vitality.

8. Untreated hive produce: beeswax; honey; propolis; bee bread.

  8.1. More beeswax is harvested.

There is so much more to say, and I'm sure you have many questions, but there is not the space here. I do hope you have enjoyed and understood this introduction to the humble bee-skep, and that you and yours, and many honeybees, might one day be appreciating some of the advantages and reasons to practice this old craft that looks to the past, and to the future...

Chris Park. Oxfordshire. 2018

3

Langstroth Hive

Tricia Nelson

I keep bees with my husband Alan. We have up to thirty colonies spread across five permanent apiaries in West Somerset. We have kept bees for 12 years and as their fascination has grown on us so have the number of hives and apiaries. We have used Modified British Nationals, WBC's, Commercials and Topbar hives.

The modern Langstroth hive bears little resemblance to the original double-walled hive designed to comply with '61 requirements deemed necessary for any hive design' as identified by the Rev Lorenzo Langstroth! That hive was not only double skinned but also double glazed around the inner broodbox!

Like many people in this country we were taught beekeeping using traditional bottom beespace, 'warm ways' Modified British National hives. Also, like many beekeepers, our colonies expanded from 2 to 6 and now to nearer 30 colonies spread across several permanent apiaries. Living where we do in the lush warm wet south west of England we rapidly discovered that most of our colonies were simply too populous for a single brood box and landed up using double brood boxes and brood and a half as a matter of routine. I hated this arrangement, always having to break the brood nest, usu-

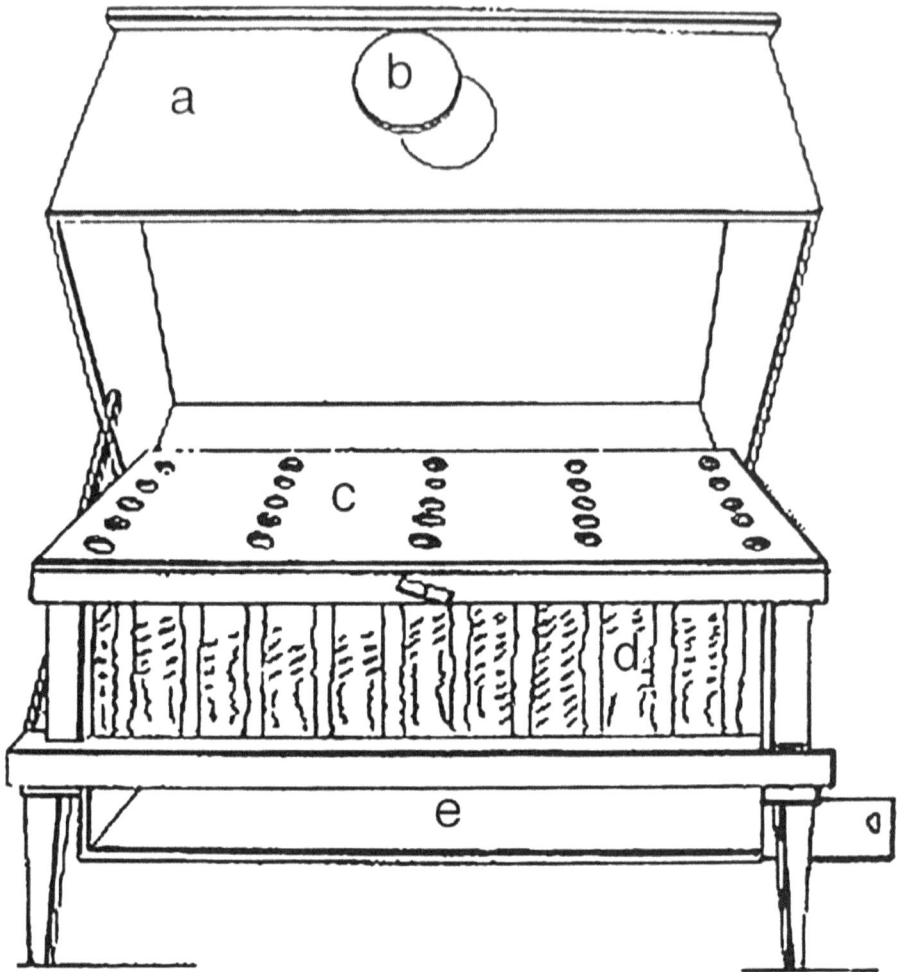

ally having to use a lot of smoke to clear bees off the top bars before rebuilding the hive. I also found any vertical manipulation of the colony, such as a Demaree or vertical swarm control, was physically very challenging because if there were any supers involved, which there invariably were, the stack was tall, not infrequently, taller than me.

Consequentially we started exploring other types of hive. We tried the Modified Commercial but with top bee space. We both liked the top bee space, which needs far less smoke to keep the bees out of harm's way but then rapidly realised that a commercial frame full of honey is extremely heavy, and so we tried other arrangements. We tried the WBC, which the bees loved. The weight problem was resolved because the internal boxes are so light and only hold 10 frames, but the bottom beespace and the need to break the brood nest remained so we explored further. Next, we made a topbar hive. Loved it. The bees loved it. However, if you want to learn about bees in any detail a topbar hive is not the easiest hive to use. If you want honey, it is also not so easy. The other problem is you are always standing to one side which makes lifting a topbar with a long comb of honey more difficult than you might think. So now we are using the Langstroth! So, let me try to describe this hive to you.

This is described as the most commonly used hive in the world, but the hive is by no means identical across the planet.

- It usually has top bee space, but not in New Zealand.

- The frame top bars are always 19 inches long.

- The frames are usually arranged 'coldways', i.e. The frames are set at right angles to the entrance.

Really nothing else about the Langstroth is finite. It is a simple rectangular box, easy to make in wood, if you have even minimal carpentry skills. It commonly holds 6, 8, 9 or 10 frames but can hold more, hence 'Long langs' (a long deep hive).

The frames can be:

- Shallows

- Mediums

- Deeps

- Jumbos

Increasingly Langstroth hives are available in polystyrene and also the heavily insulated plastic Apimaye hive.

Wooden hive and polyhive

Your hive can be made up of boxes of all the same size, or whatever size suits your bees and your requirements, or as a large brood box and smaller supers. As well as dummy boards, there are follower boards which allow you to subdivide a box into smaller units if you wish, all of which is true of most hive 'boxes'.

Another significant difference between the Langstroth hive and the Modified British National is the way the frames are designed. They are intended to be wired or left without foundation, and as a consequence there is usually no groove to support foundation in the side bars. At first sight this does seem to be a negative point but having got the hang of wiring our own frames we are now reaping the benefits. Rather than the queen leaving a line of cells unfilled where the wiring goes, she lays across the frame. When changing comb, having steamed the frames it is very quick to melt foundation onto the now pre-wired frames, and the foundation, where we use it, is cheaper! We have tried the wired Langstroth foundation and without additional wiring, other than in a shallow frame, the comb will 'flop' during inspections when newly drawn, as is the case with the majority of brood frames in larger hives.

The Langstroth deep brood box with 10 frames has a large enough capacity for most of our colonies, most of the time. This has the advantage that there are only 10 frames at most to check during an inspection, and when carrying out any manipulation which requires a second brood box the stack has not yet got too tall for me to manage.

With only 10 frames and top beespace, there is significantly less disturbance to either bees or brood, with less smoke and less risk of injury to bees, being only those on the hive walls being at risk when restacking the boxes.

Being 'coldways' takes getting used to. We spent some time reviewing our apiary layout as a consequence. Being 'coldways', the beekeeper must stand at the side of the hive, not the back in order to carry out an inspection. Many beekeepers prefer to stand at the side because they can see the entrance whilst they work. If you set your bees up in sets of 4 with each entrance facing a different way you have no choice in this. However, the thing that I find is most likely to disturb a colony's soldiers, if it is of a more defensive disposition, other than banging about, is casting shadows over the brood nest, so we generally arrange our hives in effectively an arc and always try to stand on the northern side of a hive whenever possible, which results in Langstroth hives having to face either more to the west or more to the east. The alternative if the hives are largely facing south is to stand on either one side or the other and never inspect in the middle of the day!

Langstroth frames, are wider rather than deeper, which is a shape I find easier to lift smoothly and to hold securely. They have short lugs which takes getting used to. The point being, that you don't lift the frames by the lugs, you loosen the frames and lift them by the topbar, which is the case with all frames with short lugs and should at least in theory be the case for those with long lugs also.

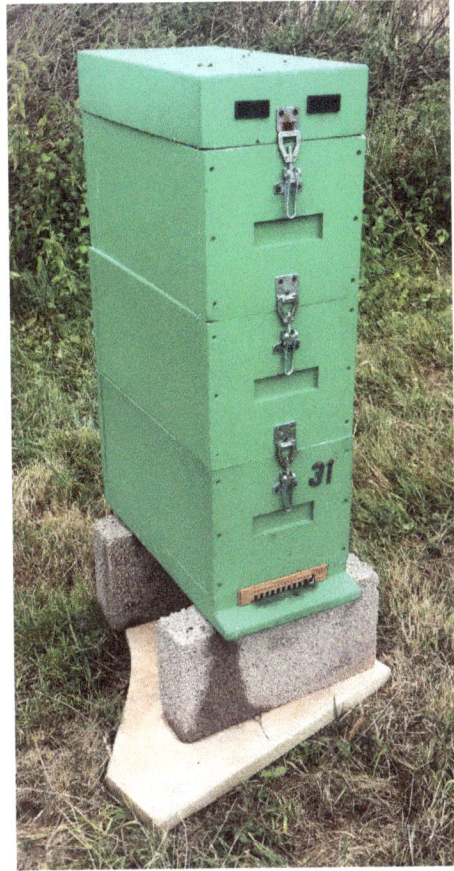

It is also possible to use Smith frames in a Langstroth if a rebate is cut in the long sides of the box. This would be a useful trick for uniting colonies from a hive using national style frames, if the lugs can be reduced in length.

At the present time we are experimenting with keeping some colonies in tall stacks of six frame polynucs, a box size which I can lift, and also 9 frame polyhives.

In conclusion I would say the Langstroth design of hive, is easy to make, versatile and practical. It is almost always the right size for our bees. It seems to suit beekeepers all around the world for whatever reasons they keep bees and with the advent of polystyrene and other insulating materials may even rival the original double-skinned, double-glazed hive of the Reverend Langstroth!

4

Layens Hive

Thornes Beehives

*D*evised by French beekeeper George de Layens in the 19th century, this hive is becoming more and more popular with thousands now in use in the USA and over 1 million in Spain alone.

Layens advocated sustainable beekeeping that rested on two principles...

1. Use local, indigenous bees.

2. Keep them in appropriate hives that require minimal management and mimic the cavity size of a hollow tree.

With big hives, big frames and plenty of room for the queen to lay uninterrupted you should be able to minimize natural swarming, produce enough honey that the hive site will allow and build up a strong colony ready for the winter.

Our hives are made from UK grown cedar from sustainable sources. The walls are 32mm thick affording far greater insulation than the average wooden hive in use today. The hive includess 14 frames (440 x 340 wide) conventionally spaced but top bars butted together at the top of the hive for added insulation.

5

WBC Hive

Bridget Knutson

I first started keeping bees in 2004, in Cheddar, Somerset. Like most people I started with one hive, went up to 13, then back to between 6 and 10 each year. I have always used WBC and National Hives. I first got interested through watching Bumble bees in the garden, did the beginners course, and got hooked. They are a very fascinating, and frustrating, insect to keep.

A bit of history first. The WBC hive (photo 1, next page) by William Broughton Carr, who gave his name to it, was first introduced in 1890 to British beekeepers. The first removable frame hive had been introduced in 1851, so it would have been quite revolutionary then. One of the first tasks the BBKA had was to try and establish a standard frame size, and a preferred hive size for the UK. They settled on the British Standard frame, and the WBC hive. This only set the inside measurements of the hive, and the outside measurements of the frames, so there was still a lot of variation. This led to difficulties for beekeepers until the British Standards for 'Beehives, frames and wax foundation' BS No.1300, came into being in 1946 and led to the British National hive and the WBC hive, and the standard frames we use today, being adopted as the national hive. The other two main types in the UK were Dadant and Langstroth, but their designs were controlled by America, and so they dropped out of fashion.

Photo 1: WBC, painted white, showing how tatty it can become unless painted regularly.

Photo 2: Unpainted lifts

The first twin walled hive was made by T C Cowan, but it was heavy, expensive and cumbersome. WB Carr decided to adapt Cowan's ideas, and make a hive which suited bees, but was still easy to use and lightweight for beekeepers. Making a double walled hive means the inside box can be lighter than single walled hives, whilst keeping the bees dry and well protected. The outside doesn't need to be heavy either, as it will slough off any water, and not penetrate the inner box.

WB Carr designed his hive to also be twin walled, but lighter to work with. He also designed it to have spacers on the frames, and designed the metal spacers himself. Nowadays with Hoffman frames, the spacers aren't needed so much, which is why there might be more propolis in a WBC than a National. The hive was designed to have bottom bee space.

I am going to concentrate on the WBC, although as I also use Nationals, there will be a bit about them as well. If you go to our beginners course, and ask what type of hive you should buy, we tell you to have a look around at meetings, ask questions, and see which one suits you. This happened to me, at the beginners course I went on, but as I wasn't going to have bees 'this year' I hadn't really had a good look, and certainly hadn't given much thought to which hive I wanted. Then my mentor told me that one of the older beekeepers in the area was downsizing, and had a hive, a suit, and some tools for sale, it seemed like a good deal, so we bought a WBC hive. The pretty one, which looks good in a garden, but that was nearly all I knew about it. I had seen one in use at an apiary meeting, but apart from thinking it looked nice, I hadn't thought about size and materials. WB Carr knew the one disadvantage of his hive was it only having 10 frames instead of 11, as the National and others do. Now, I hadn't given any thought to this, or to the size a colony can grow to. More of this later.

My first WBC was made with pine outers, or lifts, and cedar innards. I had new frames, I wouldn't reuse anyone else's frames, unless they came with the bees. I still have this, and it is still in use. Cheaper WBC lifts are made from pine, the slightly more expensive ones are made from cedar. In theory, following the British Standard, the hive is a standard size, however, the older lifts can be a slightly smaller, and may only fit on one way round when put with newer lifts. The newer ones, particularly if you buy seconds, can be slightly too large, so using them together can be interesting! I have to mix and match carefully, you really don't want a gap in the lifts so other insects can access the honey. I usually have a stack of lifts at the apiary, so if one doesn't fit very well, I can swap them around. I know I said the BS 1300 gave standardisation to hive measurements, but I think some of my lifts are before it came into play, and some are after it ceased.

The WBC is traditionally painted white, but I like the colour the cedar goes as it ages, or weathers. It starts a nice brown, and can be polished to keep that colour, but then it needs doing every year, or it can be left to weather, and turns a greyish silvery colour (photo 2) like most of mine. In these days of hiding hives from prying eyes, I like mine to blend in with the trees around, apart from the sun shining on a metal roof. I've had new roofs made for some of my WBCs, and some of them have been recovered with metal. The wood has been left under it, providing it isn't falling apart. My older lifts are made of pine, and I have found they can rot, particularly the batons which it sits on, whereas the cedar doesn't, or hasn't yet.

Because I have a mixture of old and new WBCs, I also have a mixture of painted and weathered lifts, so my hives are usually a mixture of painted and unpainted lifts.

My first bees came from a friend, who had a swarm, and we hurriedly put the hive together to house the swarm. The bees settled in, drew wax, and all was well that summer. The next year, the bees decided to swarm, as they had run out of room, I'd supered the hive, but the WBC brood box isn't big enough for a prolific queen. I then started running them with brood and a half, or, one brood box, and a super on top for extra space to the queen to lay in. This worked, but I didn't like it, so after putting up with it for several years, I looked around for ideas to change to. A WBC brood box holds 10 frames, making 45,000 cells for the queen to lay in, the National with 11 frames has 50,000 cells, the WBC 14x12 frames has 63,000 cells, and the National 14x12 has 70,000 cells with the extra frame. So changing the WBC frames from standard national brood frames to 14x12 gives the queen 50% extra laying space, much the same as putting a super on for more space, but with less hassle.

Photos 3a / 3b: Hamilton converter. The top image shows the outside; the bottom image is the inside.

By this time, I'd been keeping bees for about 6 years, so I had plenty of kit, and by that time both National and WBC hives. I looked at the choices, going to a commercial brood box meant changing the brood boxes and the brood frames, but only my national supers would fit, not my WBC supers, or moving to a deeper brood box which would take my current frames, which could be achieved with a Hamilton converter (photos 3a / 3b). So I bought enough Hamilton converters to change my current standard size brood boxes to 14x12, in effect giving my bees almost the equivalent of brood-and-a-half. I found this much easier, I don't have to worry which box the queen is in, when I separate it, or whether I would lose her if she fell out, or decided to walk out. I knew she would always be in the brood box. The frames are quite a lot deeper, and need to be handled more carefully than standard deep brood frames. The wax can easily sag on a hot day if it is spun round to check the other side. If, like me occasionally, you use the standard brood frames in a 14x12, it is even more important to handle the frames correctly, as the bees will draw comb under the frame to take it to the depth of the box. They are heavier and noticeably longer, and this is where having a hive at the right height for you is essential. I've got an old table frame set in the floor to the right height, with mesh on it, then the WBC floor, with its legs of 10 inches, to put the brood box at a height where I don't need to stoop to check it. Yes, on a good year the supers can get a bit high, but that doesn't happen every year – and they can be removed if it gets too high!

Photo 4: National brood inside WBC lifts. Note the blue edging, that is the WBC lift Gets very tight on the fingers.

I had found out, through necessity, that a National box will fit inside the lifts of a WBC, it's a bit tight, but there is still a space between the inner and outer (photo 4). I decided that this was an advantage. Although I've always been taught, and I also teach this, not to mix and match hives, I have amalgamated a WBC with a National, as it is protected by the lifts. You just need a piece of wood the length of the box, and wide enough to cover the gap (photo 5). Either tape it on underneath, or rest it on top, depending on where the WBC box is. I only do this when I want to get the bees from one hive to another, or when merging two together.

Photo 5: Merging National and WBC brood boxes. Note the bit of wood on the left hand side. This covers the gap between the National box and the WBC box. If the boxes are the other way round, I tape the wood underneath.

WBC hives are usually put with the brood frames running the warm way, but as a national brood will fit inside the WBC lifts, you can put the WBC brood box the cold way if it makes it easier for you to work it, it is just a little tight that way round (photos 6 / 7)

The WBCs have their own stand with feet, which lifts them about 10 inches up off the floor, and means water isn't absorbed beyond the legs.

The disadvantages of a WBC: it does need a bit more work to go through the bees, there is double the amount of wood to move, the lifts come off, are put to one side, then the supers are removed, but the lifts make a good stand for the supers, at a nice height to stop me stooping.

It is slightly more expensive, and if you buy it all as flat pack, then there is double the amount of work to put it all together.

If you have bees that use a lot of propolis, then the current use of Hoffman frames or castellations means your hives may be even more propolised. WB Carr designed his hive to work with spacers on the frames which if put on the runners, just lift the frames up enough to fill the gap. I do find that there can be more propolis in the WBC than the National, but it also depends on the bees.

Photos 6 / 7: Warm way / cold way – entrance is to the right of the photo. Boxes fit in either way, depending on where you want to work.

I think these disadvantages are minor compared to the advantages. I find the bees are better weatherproofed, the rain hits the outer, but doesn't get inside, the floor has feet on it, so is off the ground, or whatever I stand the hives on, less absorption of water from the base up.

I like using the lifts as a stand to put my supers on, meaning I don't have to bend so far as I would using a roof as my super stand.

There is better protection from woodpeckers, I haven't had problems with them at all, but if they did work out there is a tasty meal inside my nationals, there is only one layer of wood, against the two layers of the WBC.

Cost: There is a difference in price between the two, about £100, even buying the hives flat, and assembling them yourself. I've bought most of my kit in the sale, and some seconds, which makes it cheaper, but there is probably still a slight difference.

However, I like my WBC hives, and won't be moving away from them, even though our bee inspector calls them 'firewood'!

6

Congested Districts Board (CDB) Hive

Thomas Ellis

Thomas Ellis is a native of Co Donegal, an area with a long tradition of section honey production. He has been keeping bees for a number of years mainly uses the traditional Irish CDB. In 2013 Thomas set up Donegal Bees. He has a love for making beehives especially the CDB hive and is a great believer in the preservation of traditional beekeeping methods.

To give some history; the CDB (Congested Districts Board) was established in 1891 to assist smaller tenant farmers in the West of Ireland and parts of Scotland. Prior to this was the Land Act of 1881 which helped the larger tenant farmers but was of little use to the smaller tenant farmers, hence, the formation of the CDB. They helped the farmers by supplementing incomes. Finance was provided to boost the fishing industry by providing harbours and small piers, still present today in towns and villages across the West of Ireland. Cottage industries like carpet making and weaving were funded and eventually, attention was placed upon beekeeping.

In 1894, the CDB commissioned Abbott Brothers of Dublin to design and build a hive that could withstand the wet and windy conditions of the West. They named this hive the CDB Hive and it had to meet the requirements of the CDB for section

honey production. It was designed for black bees (*Apis mellifera mellifera*). From what I am led to believe, this hive didn't go into production until 1896.

The hive consisted of five parts; the solid floor, the brood chamber, 21 section crate, the lift/middle box, the crown board and the roof.

The solid floor

The solid floor consists of a small (4" X 4") mesh ventilation screen in the middle of the floor, with a shutter that can be opened and closed from underneath the hive; allowing access without disturbing the bees. In addition to this, there is a sloped landing/flight board which projects 5" to the front of the hive. This helps the bees to land and take off in windy conditions. The slope also prevents rainwater from running back into the hive. Where this sloped landing/flight board meets the floor of the hive – a ⅜" step was included as an extra feature to further prevent water entering the hive. This feature comes with its own complications; it creates too big a space for the bees, which stops them from climbing directly into the brood frames and it allows wind to travel up into the centre of the hive which, in turn, cools the temperature of the brood chamber; decreasing wax production. The step has since been removed from the design and the hive functions much better without it. The floor also featured four legs, 6" long that angle outwards for better stability.

Brood chamber

The brood chamber is a doubled walled box with a small porch on the front of it, which funnels rain away from the entrance of the hive. This allows rain to fall directly onto the sloped landing/flight board. The porch has a guide rail that facilitates the entrance block to slide into place to reduce or completely seal the entrance of the hive. The brood chamber overlaps the floor on both sides and the back by up to 1½" to give stability and to further waterproof the hive. The internal double wall is placed directly onto the floor, supporting the chamber. This double wall provides heat and insulation for the bees on the side elevations of the box. It houses 11 Abbott Brothers frames which are positioned the warm way in the brood chamber and are very similar in size to the standard National Hoffmann deep frames (known as DN4). The Abbott frame was distinctively recognisable by the lugs on the end of the frame and have 1½" spacing. They are known as the arrowhead frames, originating from their unique design. Behind the 11th frame, there is a single sided dummy board that is more like a division board as it is less than ⅛" from all sides of the brood chamber. This was to ensure that the bees could not pass the dummy board and gain access to the roof. It is easily removable to allow extra space to inspect the frames. It stops ⅛" away from the floor for any wind that blows through the brood chamber to escape behind the dummy board through a cone in the roof, thus, not cooling the brood chamber.

21 section crate

There are many different parts to constructing a complete section crate; the section crate and dummy board, 3 section dividers, 3 section foundation and the sections.

Irish Sections

There are many different types of sections on the market, but with the 21-section crate, Irish sections are utilised. These hold approximately 1lb of honey in the comb. The sections normally come flat packed and are thin slips of wood with V joints that are partially cut through. When these joints are soaked in water, it allows the wood to be folded into a square and is dovetail jointed on the ends to hold it together. It is important to be aware that if they are folded without the use of water – they are susceptible to breaking. When assembled, they measure 4 ¼" by 4 ¼" by 1⅞". The Irish section is then open on 3 sides to allow a 3-section length of foundation to slide into the centre of the sections. Please note, if using English sections that these differ from Irish sections as they only have 1 side split, requiring a single square of foundation per section.

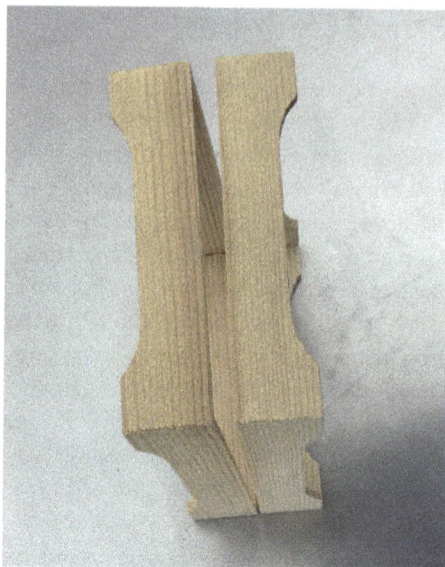

3 section foundation

It is a thin sheet of beeswax foundation, measuring 12¾" by 4⅛" and is used with Irish sections. 7 sheets are required to complete a section crate.

3 section dividers

These are a thin sheet of plastic that is placed between sections to prevent the bees from drawing the comb beyond the required width. Originally, they were made from thin sheets of wood. 6 dividers are required to complete a section crate.

21 Irish sections are needed to complete a section crate. The crate itself is a 4-sided frame with rails attached to the underside of it for the sections to rest upon. They are placed 3 rows wide and 7 rows long. Between each of the sections, a 3-section divider is placed. A dummy board is placed at the back, to tighten the sections to the front of the crate. Upon completion, a small space is created at the back of the crate, where a wedge can be placed to ensure they remain tight, which can easily be removed when the sections are full of comb. The front of the crate is easily recognisable as there is a 45° angle cut out of it on the top and bottom, to

allow the bees better access to fill the front side of the first 3 sections. The back of the crate is square and inside of this lies a dummy board that also has a 45° angled space cut out of it, top and bottom, to allow the bees additional space to fill the back side of the last 3 sections. This board also makes it easier to remove sections when they are full of honey. The crate then sits directly on top of the brood chamber, snug to the front of the hive.

The lift/middle box

The lift/middle box has various uses, having summer and winter positions. In summer position, it overlaps the brood chamber by ¾" and has the height within it to carry 2, 21 section crates. In winter position, it overlaps by 5" and still has space to carry 1 section crate within it. When the middle box is in winter position, down over the brood chamber, it double walls the front and back, whilst triple walling the sides. This provided extra heat throughout the winter months for the bees.

Crown board

It is unconfirmed if the crown board was included in the original design of the CDB Hive, but it is in use today. I incorporated this particular board into the design of the hive to make it easier to remove the section crates. Cloth sheets were used up until the crown board was introduced. The CDB crown board fits internally and has a few different aspects. It is square in size, with 2 Porter bee escapes cut out and has a bee space top and bottom. The front and back has a depth of ¾" and on both sides it has a depth of 2⅛". This is because it must be wide enough to cover the brood chamber dimensions internally and narrow enough for the section crates to sit upon, as the crates are narrower than the brood chamber.

CDB roof

The roof is the largest and heaviest part of the hive and requires over 15ft of wood to construct. The extra weight helps it to stay in place when strong winds blow, and it overlaps the middle box by ¾". The A shaped design, with the square ridge and protruding edges, ensure that any water landing on the roof, falls directly to the ground. This ridge was often made from 3" deep wood to protect the roof boards from being scored when placed on the ground. It has 2 cone bee escapes on the front of the roof to help any bees that get caught above the brood chamber after an inspection to exit the hive. Both cone escapes face outwards to prevent the bees from using it as an entrance and filling the roof with unwanted comb. The original cones were made from mesh; plastic is now used. The initial roof had a small square mesh air vent at the back of it which has since been removed as it was deemed unnecessary. Internally, the roof is large enough to carry a 3rd section crate if the middle box is in summer position.

Mechanics of CDB Hive

The CDB Hive is used today with some modifications of the original. We now use a CDB mesh floor with a correx tray underneath. It has no legs: it is placed on a hive stand. The sloped landing board remains a feature of the floor. The brood chamber is used in the same way as the original design. National DN4 frames are now used as standard, with narrow plastic end spacers on the lugs to prevent the bees gaining access to the roof of the hive. Previously, the Abbott Brothers frame had the spacer built into the lug. The first design of the hive did not include a queen excluder and it was discovered, with section honey, the queen rarely went to lay in the crate. This is partly due to the fact that the sections are a different width to the deep frames with the spacers below. In a National Hive, the brood chamber and super directly align on top of each other and the frames in both the brood chamber and super also align the same way as they are the same width, creating a need for a queen excluder. With the sections placed on top of the CDB brood chamber, this offsets the line between the two, making it increasingly difficult for the queen to go directly up, discouraging her from laying in the sections. As a further prevention, a flat slip of cardboard is used across the middle 3 frames of the brood chamber (5th, 6th & 7th frame) and this stops the queen from laying at the highest point in the brood chamber. This also encourages the bees to fill the front and back sections of the crate. In a season, it is common for bees to eat through the cardboard, needing replacing for the next season. It is well known in section honey production that the bees only fill the middle sections in a section crate, hence, with this cardboard in place, it encourages them to fill the crate more evenly as they don't have direct access to the middle sections. When the crate sits directly on to the brood chamber, ensuring the cardboard slip is in place, they must be kept tight to the front of the hive. The middle box then sits down over the crate leaving a cavity on the sides and back of the box. It is then filled with breathable insulation as the more heat provided, the better the sections and the faster they will produce wax and fill them with honey. On top of the last crate, a flat cloth sheet is placed, and the crown board is then placed on top of this. The roof is the final piece to be put in place when assembling the hive. It can often happen that the bees don't perfectly fill the individual sections. To get the maximum amount of perfect sections, the completed ones are taken out of the crate. Two crates are then used; one with ¾ filled sections and one with sections that are ½ filled or less. The ¾ filled section crates are put directly on top of the brood chamber, followed by the crown board and finally the ½ filled section crate is put on top of this. The bees tend to dislike anything between them and their honey, so they remove the honey from the crate above, take it down through the crown board and finish filling the ¾ filled section crate. It is important to know, the later in the season this is completed, the more honey lost to the brood chamber. When the sections are removed, any brace comb and propolis is removed from them. They are then placed into cardboard section cases for storage and for sale at a later date.

To conclude, the CDB Hive is by far the best hive to produce sections in. The design is well thought out and planned to the smallest detail taking into account all possible aspects.

7

British Modified Commercial Hive

Chris Utting

I started my beekeeping in 1983 in Westward Ho!, North Devon with Modified National hives. I wanted honey to sell to the tourists and my hobby grew into a business with up to 30 colonies. But to get lots of honey you need lots of bees. I then worked with a National brood and a half and then a double brood system. This enabled the larger colonies to develop. But checking through 22 brood frames every week was hard work. Then I discovered the Commercial brood box with only 10 brood frames to check. The changeover was obvious. The National floors, queen excluders etc. are were compatible. As time passes, I am now down to 6 colonies.

It was in 1983 that I started to actually handle bees and hives at the branch apiary. So far as me, my fellow students and our teacher were concerned the only hive that we used was the Modified National. This was the hive that everybody else used. Everybody in my small world of beekeeping in North Devon appeared to use the National. There was no other hive. So, I conformed.

As my beekeeping experience developed and I started to increase the number of colonies I soon became aware that the National brood box of 11 frames was not big enough. To get more honey I needed more bees. I wanted to produce as much as honey as possible to help pay for this hobby. So, I developed my methods moving up to a brood and a half and then to a double brood. This worked and I was getting the increased bees and potential honey crop.

Then I realised how much time it took carrying out a regular swarm check when I had two boxes to examine and up to 22 frames to check.

Soon I discovered the Commercial hive. A single walled hive with outside dimensions 465mm x 465mm; only 5mm greater than the National hive with brood frames 421mm x 440mm (16" x 10") and with a bottom bee space.

In his 1904 book "£300 per annum from 30 acres" Samuel Simmins in putting the case for the larger brood frame also gives an interesting criticism of the BBKA.

"I must state without hesitation that the Standard frame of the British Beekeepers' Association is much too small for any bee-keeper who is attempting to produce honey on a wholesale scale. It is true I have been using the Association Standard frame largely for some years past, and expect to continue to do so as long as I supply bees to those who have adopted that size; but its use has only the more forcibly brought to my mind the decidedly superior advantages enjoyed when using a frame measuring 16 inches by 10 inches.

Evidence in favour of the larger size, as giving greater security in winter; a larger population more rapidly developed in spring; less inclination to swarm; and at all times a more prosperous and profitable colony, with comparatively little trouble in maintaining that prosperity has been accumulating right along, as shown by the practical results secured from such colonies as remained in the old frames used by myself and other apiarists, and which should have been, and will yet be recognised as the Standard frame in this country, viz. :— 16-in. by 10-in.

It does not denote progress to hold to a certain size of frame simply because that has once been stamped as the Standard of the British Beekeepers' Association, whose committee, because of its peculiarly exclusive organisation, was not in a position to deal "understandingly" with the commercial interests of either the reproducing or manufacturing industry. There has long been a smothered feeling of opposition in some quarters against the frame now in use, and yet there are many who refrain from raising a dissenting voice for fear of coming into conflict with a recognized institution which has not one valuable point as a recommendation, nor a single excuse for its continued existence, seeing that every honey producer using the small frame is at a serious yearly loss, as he may soon ascertain on trial."

So, I had not discovered a new design of hive. It has been around probably since late Victorian times.

The advantages were obvious as the volume of the brood box was so much bigger than the National brood box. Equivalent to a National brood and a half but only 10 or 11 frames to check. An instant reduction in the time spent on swarm inspections. The Commercial brood box consisted of only four pieces of timber compared to the National with eight pieces and was much easier to assemble. There is a seven-minute film in YouTube that demonstrates how to assemble a brood box. Search for 'Thorne Beehives Commercial'.

I gradually changed my hives.

I found that the Commercial floor, queen excluder, super, crown board and roof were of little significant difference in the length and breadth dimensions there being a difference of only a few millimetres compared to the National equipment so I only needed to obtain the Commercial brood box. There are two choices of Commercial frames for the supers. Either with the Hoffman side bar spacers or the straight edge Manley side bar being 6mm wider making the uncapping easier but check that the shallow frames, being longer than the National, will fit into your extractor. As a comparison the number of worker cells in the brood box is 70,500 to the National of 50,000.

However, I realised that the best way of generating income was not the sale of honey but the sale of nuclei to beginners. A quick look at the potential market and I found that the beginners were at the same state as I was a several years ago. They all wanted to use National hives so as I was producing nuclei for sale I retained some Nationals for my customers. As I was working both Commercial hives for my bees and National hives for the beginners I discovered a very useful accessory in the form of the Hamilton converter. By placing this four-piece frame on top of the National brood box I could insert ten Commercial brood frames.

There are some disadvantages of the Commercial hive. I have been a beekeeper for

nearly 40 years and am becoming aware of the heavy weight of a full brood box. I have recently overcome this by using an empty Commercial brood box to temporally hold half the brood frames thus reducing the weight to manageable amount. Also, when lifting the brood box there are only fingertip D shaped recesses on the four sides. I have overcome this by attaching batten handles on either side to get a better grip. The addition of these grips will increase the width of the box if you are stacking them on to a trailer. Woodpeckers are uncannily attracted to the recessed finger holds as it requires only 50% of the pecking effort to penetrate through to the brood area.

The formed metal runners leave a small space to the side of the box and some colonies that collect more propolis than normal can fill this gap thus making manipulation of the brood frame more difficult. This can be overcome by fitting a flat metal strip 25mm deep and 1.5mm thickness to leave a wider gap.

For those beekeepers who prefer to use thick gloves the smaller lugs on the top bars of the brood frames may be more difficult to grasp.

Advantages	Disadvantages
Only 10 or eleven frames to check saves time when looking for swarming signs	Heavy brood box when full
Four-piece simple construction	Fingertip handling recesses
Construction video on YouTube	Added handles increase storage width
Cold or warm way orientation	Recess attract woodpeckers
More space for a large colony	Formed metal runners attract propolis
Hamilton Converter of National brood boxes provides management options	Narrow lugs of brood frames harder to grip
Hive parts interchangeable with National hive parts	
Bottom bee space	

8

Modified Dadant

Richard Ball

I have not owned this hive myself but in my career as a bee inspector I manipulated Modified Dadant and Buckfast Dadant hives hundreds of times. They are a good hive to use provided you have assistance to lift full supers, prolific bees and your apiary is in an area with a plentiful nectar flow.

With the publication of Langstroth's design for a moveable frame hive the concept of modern hives was born and many hive designs then proliferated in the U.S.A. and elsewhere around the world. The Dadants of Illinois were and are a well-known bee-keeping family setting up the largest bee supply company in the world. The Dadants designed hives using 10 frames measuring 18½ x 11¼ inches developed by Moses Quinby. However, in 1917 they introduced a hive using 11 17⅝ x 11¼ inch frames being the Langtroth width with the Quinby depth. It was called the Modified Dadant or M.D. and is the hive we know today. It is the largest beehive generally obtainable throughout the world and was very popular between the two world wars. It was used and promoted by R. O. B. Manley, inventor of the 'Manley' self spacing super frame, who also wrote several very good beekeeping books including 'Bee Farming'. Brother Adam at Buckfast Abbey used a modified version holding 12 frames making it a square hive known as the Buckfast Dadant. The Modified Dadant hive has also been popular in continental Europe though there are often slight differences in sizes due to metrication. This popularity may have been the result of the founding father of Dadants, Charles Dadant, having emigrated to the U.S.A from France in 1863.

They are readily available commercially made, assembled or in the flat (in the flat means that you buy the ready-made wooden components to glue and nail together yourself). A reasonably competent carpenter can make his or her own from plans, which have been published in the past. Materials used include western red cedar, pine or nowadays various plastics. Western red cedar is a better choice than pine because, though more expensive, it is lighter, rot resistant, less inclined to warp, doesn't absorb as much moisture and lasts a lot longer. I have used Western red cedar hives for over thirty-five years, and they are still serviceable. When assembling or buying second hand make sure that the components are square by checking that the diagonal dimensions are equal.

The hive

Comprises of various components described below. There is no recognized standard specification for this hive, but the dimensions given here ensure compatibility with the leading British producers and have been taken from UK government promotional literature. (MAFF Leaflet L549, long out of print) Imperial sizes are given, as this is the standard in the U.S.A. and used by MAFF.

A floor

This is 20 inches x 18½ inches x 2 inches deep with an entrance on the narrow side. The floorboards are ¾ of an inch thick and are lapped into the ⅞ inch thick side rails. They are set so that one side of the floor leaves a bee entrance of ⅞ of an inch one side and ⅜ of an inch on the other. The reason for two sizes is that the deep side is used when foundation is placed in the brood box, e.g. hiving a swarm, shook swarm or creating an artificial swarm, in this way the bees draw the foundation to the bottom of the frames. If the narrow entrance is used, then the bees chew away the bottom of the foundation. When the comb is completely drawn the narrow side is then used which stops the bees drawing comb from the bottom bars and fixing it to the floor. Entrance blocks can be used with the wider floor entrances to reduce the entrance size or to close it. Wooden slips can be made for use with the narrow entrance or foam strips can be used to close or reduce the entrance when using either width. When locating the hive, it is sensible to slightly slope the floor forwards to help drain water off the floor if there is any driving rain.

With the advent of *Varroa* and the use of open mesh or other designs of floor it is better to use a ⅜ inch deep entrance as with wider entrances the bees build comb between the bottom bars of the frames and the mesh floor making it difficult to remove the frames.

Brood Boxes and Supers

These are also 20 inches x 18½ inches with the walls made of ⅞ inch finished thickness timber. Brood boxes are 11¾ inches deep. Supers are 6⅝ inches deep. The walls should be

lock jointed together. A rebate is cut into the inside face of the narrower end walls 7/16 of an inch into the thickness and 7/8 inches deep if runners are used, 5/8 inches deep if not. D shape finger holds, 3½ inches long, ½ deep and 2½ inches from the top of the box are cut into opposite walls to aid lifting.

Eleven Hoffman self-spacing frames fit into the brood box. They are 17⅝ inches long and 11¼ inches deep with short lugs and spaced at 1½ inch centers. This is slightly wider than the Langstroth at 1⅜ inch centers. Brother Adam and many other beekeepers used studs or screws to achieve this spacing instead of using Hoffman spacing in order to reduce propolising at the contact points.

Usually 11 Hoffman or 10 Manley self-spacing frames are used in the super at 1½ inch or 1⅝ inch centers respectively.

Section racks

Also being 20 x 18½ inches but only 4 5/16 inches deep. The Rack is designed to contain 40 4¼ inch square sections. They are commercially available.

Crown boards

Originally thin wooden boards were used but today plywood is used with feedholes cut into the centre. Because this is a top bee space hive 7/8 x 3/8 inch battens frame the upper side of the board only.

Roofs

The roof is flat so that when inverted the hive with screen board attached can be placed inside it and slid over flat beds or car boot floors. They also make a handy support for supers and brood boxes when manipulating colonies. Roofs are 6 inches deep with an internal size of 20⅝ inches long and 19⅛ inches wide.

Queen excluders

These are 20 inches long and 18½ wide. Wire excluders should be used with a batten surround on the top surface to ensure a bee space under the super. Zinc queen excluders of this size are sometimes found second hand. It is best not to use them but if you do, they will need support with ⅞ inch x ¼ inch wooden battens, around the sides and three battens equidistant across the length. They are used with the battens uppermost to maintain the correct bee spaces.

Also designed are five frame nucleus boxes, which are half the width of a standard hive. They take 5 Hoffman self-spacing frames. Two can be placed over or under a brood box or super, handy for many beekeeping applications.

To improve their longevity wooden hives should be protected with a non-insecticidal preservative. It is best not to use paint, as when moisture gets into the timber it will blister and become unsightly.

In use

This hive is the largest in production throughout the world, with a total brood frame area of 2,181 square inches. Many beekeepers say it is too large but are happy to keep bees in double brood box Modified National hives, which equate to 2,618 square inches! This hive has mostly been used by commercial beekeepers who covet its large capacity, reducing the propensity to swarm, and the large super capacity resulting in fewer units to move around and extract. Its popularity is in decline so why is this when great beekeepers like Manley and Brother Adam liked and promoted it so much? If we go back to 1882 when the BBKA settled on recommending the British Standard frame the majority of bees kept were the native black bees. This bee had been selected over centuries to suit skep beekeeping and was suited to the small standard frame. Brother Adam writing in his book 'Beekeeping at Buckfast Abbey' remarks that the small framed hives used at the Abbey were quite adequate for the British black bee. However, between 1905 and 1919 during the Great Epidemic the world was subject to a period of severe bee losses that effectively wiped out the black bee in the U. K. As a result, many bees were imported to make up for the losses mostly from Italy. They were far more prolific and not suited to the small British frame. At this time Brother Adam started to use a Dadant type hive, but before committing to a total change he compared the results of using prolific queens in double brood box British Standard framed hives compared with the single Dadant type brood box. Honey production was greater with the Dadant type, so he committed to his own twelve-frame version known as the Buckfast Dadant.

With prolific bees, large brood boxes are needed because a queen has the ability to lay over 2,000 eggs a day. This means that some 42,000 cells are needed to accommodate the developing brood and leave 2,000 hatched cells a day for egg laying. This equates to 1,680 square inches of brood comb and allowing some extra space for stores makes the Modified Dadant an ideal hive for prolific bees. It will also hold 70 lbs of stores, which are needed to overwinter such bees. The Commercial, Jumbo Langstroth and 14 x 12 also offer a similar capacity.

Another criticism often leveled at the Modified Dadant is that it is too heavy but is this fair? I think not, as any of the larger hives are just as heavy to move and will require two of you. Yes, the brood frames are heavier than most especially the British Standard and if you are used to the latter can be a bit of a surprise, but by holding the top bars rather than the frame lugs they are easy to lift, manipulate and examine. A super full of honey weighs 43 lbs rather than the 30 lbs of a Modified National so assistance is useful whichever hive you use. However, because of the greater capacity there will not be as many supers on a hive.

Bee farmers rather than hobbyists have used this hive; in particular Brother Adam, who had lots of help, and Manley, who with lower wages had apiary inspections teams of two apiarists and a smoker boy. However, its popularity is declining why is this? Difficult to say, perhaps it just doesn't have a champion today. With all the current disease issues perhaps, the bees are not so fecund, so larger brood boxes aren't necessary. Maybe it is changing agricultural practices, as when Manley and Brother Adam were active between the World Wars wildflower meadows managed for hay was normal forage. Today we see rye grass drowning out wildflowers and cut three times a season for intensive silage production and hedgerows slashed back reducing forage for bees. I know this as the 'Green Desert'. To day with Commercial beekeeping being mostly rural based perhaps the green desert syndrome is the reason, but this hive is good and has its place with prolific bees and abundant forage. Today this is often found in city and urban areas.

In summary for prolific bees and forage the Modified Dadant hive is an ideal choice.

Richard Ball, Chairman DARG

Advantages	Disadvantages
Single wall	Heavy to move
Simple construction	Large frames can appear heavy to manipulate
Top bee space	Supers are heavy when full
Ideal for prolific bees	Not readily available second hand
Large capacity helping to reduce the swarming impulse	Not really practical without apiary assistance when supers are full
Adequate winter storage space	
Larger capacity super means fewer are used	

9

Smith Hive

Willy Robson

Willy Robson keeps 1500 hives of black bees in the Scottish borders. He has kept bees for fifty-five years and has always used Smith hives. His family have kept bees since at least 1900 in these parts.

The Smith hive was designed and built by W.W. Smith of Innerleithen, Peebleshire in the Scottish Borders. He kept 120 very powerful colonies of black bees for heather honey production. He read many books about beekeeping written by American authors and decided to build a hive on the American pattern but to the British Standard size. The dimensions were 18¼ by 16⅜ in. to accommodate 11 BS deep frames (short lugs 15½ in.) using ⅞th in. timber all round with a frame spacing of 11⁵⁄₃₂ inch and a half was too tight. The brood bodies were made of four pieces of wood half lapped at the joints and were cheap and simple to make – much more so than the National.

The floors could be extended at the front or otherwise flush down the front and were ⅜ths deep to prevent mice getting in (full width entrance). The crown boards were flat on the bottom side and with a shallow rim on top (top bee space) and often incor-

porated a swivel to provide an entrance for a brood body containing a nucleus on top. A feed hole was provided with a cover. The roof was constructed to fit over the brood body with suitable clearance all round and could be up to 10 in. deep and covered in flat galvanised iron or aluminium folded at the corners. The shallow boxes were 5⅞ in. deep, again made from ⅞th in. timber and would usefully accommodate 10 Manley-type shallows 1⁵⁄₈th in. spacing for honey production.

R.O.B. Manley designed frames with closed ends right to the bottom bar that didn't swing about in transit. They also double-walled the super on the two long sides which was important on heather sites. The hive would normally be constructed in western red cedar. Willie Smith built his hives in white pine which, once it was dry, was very stable. He also painted the hives in white lead paint which looked well. This was a cheap and simple hive to make in large quantities. The worst feature of the Smith hive was the handholds which were very often too shallow to get a grip.

Willie Smith was well known to our family and we learnt a great deal from him. He built his colonies up by giving the bees several deep boxes from which heather honey had been scraped and which still contained a good amount of honey. Thus they became very strong to be ready for the bell heather in July and the ling in August. The bees stayed in the same sites winter and summer and were able to fly to the heather without being moved. This was a very late area and most of the queens were superseded during August, sometimes when they had reached four years old. Mr. Smith ran a very successful business.

He kept the bee space on top according to the American pattern so he could tip up adjoining brood boxes without dislodging the frames in the bottom box because of brace comb. This might not be the case with a bottom bee space.

10

Modified National Hive

Richard Ball

In the 1970s I became interested in self-sufficiency so purchased a cottage with a field and small wood, grew my own vegetables, kept goats and managed my own firewood. I was particularly drawn to the concept of beekeeping with regard to pollination and producing a few jars of honey. After some training with a semi commercial beekeeper I started to keep them in 1982 and ran up to 20 colonies migrating most colonies through the season to Oilseed Rape, Field Beans and Heather. During 16 years as an authorised bee inspector I saw thousands of colonies in most types of hives and other containers. With retirement I have stopped migrating bees and run about half a dozen colonies.

During the 19th Century the concept of modern hives was developed in the U.S.A. with the publication of Langstroth's design for a moveable frame hive including the bee space around the frames. This bee space was the breakthrough needed to design effective beehives. Many hive designs proliferated not least in the U.K. so in 1882 the British Beekeepers Association met to decide on a standardized brood frame size. They settled for a brood frame 14 inches long and 8½ inches deep with long 1½ inch lugs known as a British Standard frame. A super frame of the same length and 5½ inches depth was also agreed. These frames then formed the basis for a plethora of hives including the development of the WBC hive by W. B. Carr which went on to become the most used hive in the UK. However its use has declined in popularity over the past 60 years. These hives had many components and did not necessarily have interchangeable parts between mak-

DN1 frame fitted with a plastic end spacer, a DN4 Hoffman frame as recommended by BBKA and a wide bar DN5 frame with the shoulders removed and stud spacing fitted.

ers; so practical beekeepers sought a simpler design of hive using single walls and the British Standard frame.

In about 1920 the problem of using a long lugged frame with a single wall was solved by making the end walls double and letting the frame lugs sit on a lower internal end wall. One pattern of this type known as the Simplicity Hive, became very popular. It certainly lived up to its name in comparison to the WBC, then advocated by the BBKA, as it was a series of 18⅛ inch square boxes with a flat floor and flat roof. Each box containing 11 British Standard frames and they could be placed warm way or cold way without having different floors.

In the USA there was another hive known as the Simplicity. This created a lot of confusion so the British Simplicity became known as the National. These hives are sometimes seen on the secondhand market today.

Later the double end walls were changed using a rebated 1 x 1½ inch bar that was attached to the inner wall. This version of the hive became known as the Modified National and was included in the British Standard 1300 of 1946 and also in subsequent editions. Its advantages over the earlier National are that it uses less timber to make and the external framing for the frame lugs make very good hand holds in comparison to the 'D' cut outs used in other single walled hives.

They are readily available commercially made, assembled or in the flat. (In the flat means that you buy the ready made wooden components to glue and nail together yourself) A reasonably competent carpenter can make their own from plans, which have been published in the past. Materials used include western red cedar, pine or nowadays various plastics. Western red cedar is a better choice than pine because, though more expensive, it is lighter, rot resistant, less inclined to warp, doesn't absorb as much moisture and lasts a lot longer. I have used western red cedar Modified Nationals for over thirty-five years and they are still serviceable. When assembling or buying second hand make sure that the components are square by checking that the diagonal dimensions are equal.

The hive

Comprises of various components described below. Imperial sizes are given as they were originally used and the metric equivalent may not be exact.

A floor

This is 18⅛ inches square x 2 inches deep with an entrance on one side. The floorboards are ¾ of an inch thick and are lapped into the ¾ inch thick side rails. They are set so that one side of the floor leaves a bee entrance of ⅞ of an inch one side and ⅜ of an inch on the other. The reason for two sizes is that the deep side is used when foundation is placed in the brood box, e.g. hiving a swarm, shook swarm or creating an artificial swarm, in this way the bees draw the foundation to the bottom of the frames. If the narrow entrance is used then the bees chew away the bottom of the foundation. When the comb is completely drawn the narrow side is then used which stops the bees drawing comb from the bottom bars and fixing it to the floor. Some manufacturers make a sloping floor giving a wide entrance on both sides; this helps to drain water off the floor if there is driving rain. With the flat floor this can be achieved by tilting the hive slightly forward. Entrance blocks can be used with the wider floor entrances to reduce the entrance size or to close it. Wooden slips can be made for use with the narrow entrance. Foam strips can be used to close the entrance when using either width.

With the advent of *Varroa* and the use of open mesh or other designs of floor it is better to use the ⅜ inch deep entrance as with wider entrances the bees build comb between the bottom bars of the frames and the mesh floor making it difficult to remove the frames.

Floors are sometimes seen with extended fronts to make a landing board. These are fun to observe the bees but they are not necessary and are awkward when moving colonies, as they will not fit into the roof.

Brood Boxes and Supers

These are also 18⅛ inches square with the walls made of ¾ inch finished thickness. Brood boxes are 8⅞ inches deep. Supers are 5⅞ inches deep. A cross section of the frame end walls is shown. These boxes are designed and usually built as bottom bee space but can be supplied commercially or altered to be top bee space.

Eleven frames fit into the brood box. It was designed to take eleven DN1 frames with metal ends, plastic nowadays, to space the frames at 1⁹⁄₂₀ inches or 37mm centers but today it is recommended to use Hoffman self spacing frames at 34mm centers. With this narrower spacing a dummy board should be inserted at one end. It is possible to hang 12 Hoffman frames in a box but is not good practice as when they become propolised it can be difficult to take out the first frame and this manipulation can put extra strain on the boxes integrity. Various frame types can be used in the supers details being found in text books and equipment catalogues but the broad rule is to hang in them 9 drawn combs or 11 frames of foundations. When using the British Standard frames the brood box spacers are used for 11 combs and a 2-inch spacer is used for 9. Castellated frame runners are available with 9, 10 or 11 slots. These are helpful as when moving supers around the fitted frames remain securely in position. 9 and 11 slots are used as above but 10 slots is a compromise; and with a good nectar flow fresh foundation is usually drawn out without brace comb. Castellated runners should never be used in brood boxes.

Supers combs held in place by castellated runners.

Section Racks

Also 18⅛ x 18⅛ inches square by 4 ⁵⁄₁₆ inches deep. The Rack is designed to contain 32 4¼ inch square sections. They are commercially available. If you only want few sections special frames are available for use in a standard super alongside your chosen super frames.

Crown boards

Originally thin wooden boards were used but today plywood is used with feedholes cut into the centre. Because bottom bee space is used ¾ x ⅜ inch battens frame each side of the board to create a bee space over the frames in the upper box.

Roofs

The original and recommended roofs are flat so that when inverted the hive with screen board attached can be placed inside it and slid over flat beds or car boot floors. They also make a handy support for supers and brood boxes when manipulating colonies. Roofs come in various depths, generally 4, 6 and 8 inches. Deeper roofs are less likely to blow off in strong winds so use 6 inches or greater. Gabled roofs are available which give a pleasant 'old world' look in the garden but are not as practical in use especially when moving colonies.

Queen excluders

18⅛ inch square queen excluders are used between the brood boxes and supers or the floor and brood box when wishing to prevent absconding. With bottom bee space zinc, metal, or plastic queen excluders can be used but it is better to use framed wire excluders. With top bee space only wire excluders should be used.

Also designed are five frame nucleus boxes which are 9 ¹⁄₁₆ inch wide so side by side they become 18⅛ inches square. Therefore they can be placed over a brood box or super. This is very useful when drawing new brood comb as each of these boxes will be half the weight of a brood box full of honey, which is about 27 kilos or 60 lbs.

To improve their longevity wooden hives should be protected with a non-insecticidal preservative. It is best not to use paint, as when moisture gets into the timber it will blister and become unsightly.

In use

This hive is very adaptable to beekeepers needs because of its single walls, shape and relatively compact size. However there is no doubt that using a single brood box makes skillful management necessary with a prolific strain of bee.

When managing a single brood box in spring it is important to ensure that there is no congestion in the brood box by too many stores. This is true with all hives but is more pronounced in the smaller types. In the late summer, after removal of the honey, immediately make sure the bees have sufficient stores until the next inspection. Subsequently assess the stores present and feed sufficiently to ensure the bees can overwinter without an excess to congest the brood area in the following spring.

It cannot be stressed enough that for successful management of this hive a plentiful supply of clean drawn brood comb is needed. The best way of ensuring this is to use a few brood boxes fitted with frames of worker foundation used as supers. When full of honey these frames can be extracted and stored over winter ready for the new season. Most extractors with tangential cages or screens are suitable for this. These combs make it easy to remove damaged or congested combs on a nice day in late March before the brood nest reaches the sidewalls. It is not possible to use frames of foundation at this time because it is too early for them to be successfully drawn and the bees will spoil the foundation. Waiting till late April or May to insert foundation will usually be too late as the brood will be on every frame. Drawn comb

Bailey comb change being applied to a Modified National. Note the shallow eke between the brood boxes to make an entrance between them.

is essential for use with the Demarée method of expanding the brood nest and can also be helpful with artificial swarming.

The flexibility of this hive enables beekeepers with prolific bees to enlarge the brood nest by stacking two brood boxes (double brood) or a brood box and a super (brood and a half). These systems require slightly different management techniques and have the disadvantage that there are double the number of brood frames to examine. However when carrying out examinations for swarm control the two boxes can be split open, lifting the top box like a hinged lid, and looking at the bottoms of the top frames. If queen cells are present a full examination needs to be carried out, if not, it is unlikely that there will be queen cells elsewhere.

The comparative smallness of this hive also makes Snelgrove manipulation easy or other methods of artificial swarming such as Pagden by placing the parent colony facing in the opposite direction over the artificial swarm colony. This is handy if space and spare roofs are not available!

If the beekeeper wishes to use a larger single brood box then a 14 x 12 inch brood box is available. This is about half as big again so roughly equivalent to a brood and a half. Also there is an adapter, or eke, to fit a standard Modified National Brood Box converting it to hold Commercial Hive brood frames.

Well used Modified National hive.

In Single Brood box management systems a unit of floor, brood box, bees, excluder, empty super and travel screen is moderately easy to move by one person, so is ideal for migratory beekeeping; with the larger hives this is definitely a two person job.

One problem is that the pollen arch over the brood is extended into the bottom of the central super combs over the queen excluder. It is best to mark these supers and frames so they are returned to the same place in the following season.

Modified National set up to be transported in a private car. Note the hive is sitting in its roof and the ventilation or screen board fitted to the top

Advantages	Disadvantages
Single wall	Bottom bee space
Easily handled brood boxes and supers	A single brood box is a little too small for prolific queens early and mid season
Brood comb can be readily drawn in a brood box used as super	First super combs become clogged with pollen
Readily available new or second hand	More complicated construction than some other single wall hives
Components are interchangeable with the National Hive	
Easy to use for migratory beekeeping	

11

14x12 or Deep National Hive

Glyn Davies

I have only kept bees in South Devon, just a few miles from Buckfast. Once I kept up to 40 colonies but now I am reduced to around 12! Beekeeping was an ambition from childhood inspired by a primary school visit from a lady beekeeper. I should qualify for BBKA 50-year certificate in 4 years's time! I've used most conventional or near conventional hives starting with Commercials, but now the 14x12 is my favourite.

For all its conveniences and popularity, you don't have to have been a beekeeper for long, just about as soon as your first beginners' training session, before you had been told, "*The standard Modified National hive is too small for modern, highly productive queens.*" And you are not long into the practical craft before you realise exactly what that means.

The first super over the brood chamber hopefully for honey storage, will be packed with pollen top and sides, and the centre of the box with empty polished cells which the bees have provided for their industrious, fecund queen. But they, you and the queen are frustrated as she is deliberately excluded. Not only does this limit the colony productivity but the frustration of lack of laying space could encourage swarming.

Woodland setting for 14x12.

Frustrated and imaginative beekeepers in the past lifted the excluder above the first super which satisfied the bees but doubled the work for the beekeeper who now had to look through 22 combs to find the queen instead of the basic eleven. Amazingly it was not until around 1980 that someone whose identity has been lost had the bright idea of just extending the length of the national side bars by three and a half inches and the depth of the Brood chamber to take them. Thus, came into existence the increasingly popular 14 x 12 hive. All other dimensions are identical with the National hive, so this was an economical as well as a commonsense alteration.

Of course, it does make brood chambers, frames and sheets of foundation more expensive than the basic National but not when the cost of the super is included that is used in the commonly called "brood and a half" system.

The cost of a change to 14 x 12 from National is made easier and less expensive by attaching a simple 3 ½" eke to the bottom edge of a national brood chamber. A simple, strong clip system is available. Two are needed on opposite sides and the combination is temporarily, and very securely held together by a metal sliding wedge hammer-tapped tightly into position. This is seen in the photograph. A cheaper and very effective attachment can be used using a trio of three-hole plates, two on one side and one opposite. Again, see the relevant photo.

National and 14x12 frame size comparison.

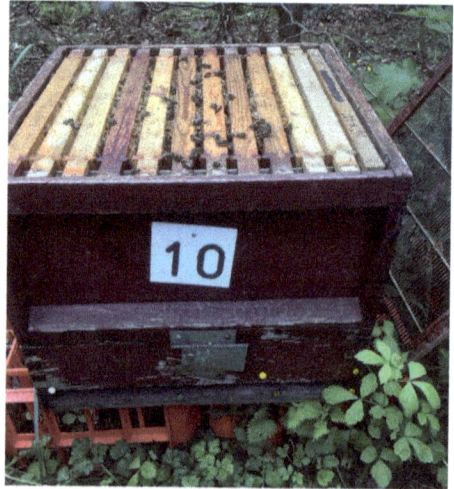

Adapted Modified National.

The bees seem to like the available frame area which allows a match more closely to their natural, approximately spherical early-season brood nest. This is noticeable when lifting a central 14 x 12 frame which being a vertical section through the nest is often seen as a perfect circle. Other frames are more landscape rectangular and just don't fit the instinctive and physical egg laying habits of a queen in continuous laying mode. She has to spend less time searching for empty cells. Altogether a more efficient and natural system.

The 14 x 12 frame has the long lugs of the National. This is a good feature especially for novices who are getting used to handling frames of live bees with the necessary gentleness

Easily handled frames.

Above: Room for the entire brood nest and stores of honey and pollen.

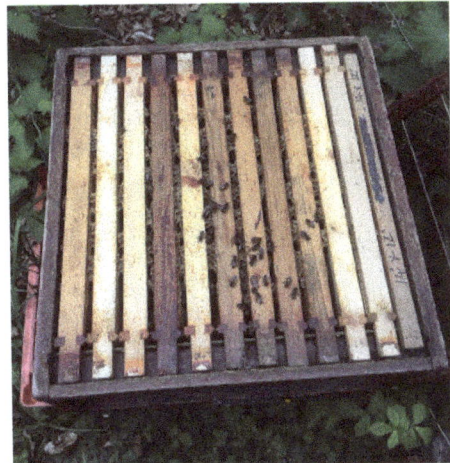

Right: 11 frames plus dummy board in full use.

and dexterity. It is noticeable, especially in a frame with plenty of stores, that the centre of gravity and hence balance of a frame when handling it is very different from other frames. Correct handling is vital. Flipping a frame to examine both sides can cause a disaster especially if it's heavy and the weather hot. Wired foundation is vital but the additional length of wire and the habit of bees not to attach the comb to the bottom bars does mean more gentle handling and correct frame turning. Not a bad discipline. Wiring the frames rather than using embedded wired foundation is probably a better system but the convenience of wired foundation gets my vote. A trick I find useful is to have pocketed in my bee wear a small container of drawing pins and cocktail sticks. The latter to examine suspect larvae and the pins for a variety of needs but in this context, they are useful to pin the exposed wire loops to the frame bottom bars. No worries after.

A 14 x 12 frame is very heavy when containing a good supply of stores and the long lugs help lifting. Moving a full brood chamber, even for perhaps a Pagden-type artificial swarm manoeuvre definitely is best with two people and the combined extra height and weight make a Demarée system or Bailey comb change a real challenge for a short person working alone. In the autumn the weight of stored honey in the brood chamber is very reassuring. Often feeding for winter is unnecessary. An attempted heft will release a large grunt of effort, even before the Ivy flow.

To take full advantage of the space within the 14 x 12 brood chamber it is necessary to have a good quality queen. She must be healthy, well-fed as a larva and mated in good weather with a rich variety of drones in her selected congregation area. Her colony must be able to build up quickly in the spring. This is especially true if a spring crop of honey is expected. The first super will be generally added later than with a single national brood chamber but with a large foraging force coming along, full advantage of apple, dandelion and OSR will be possible with no half empty, first super. When supers are removed for a spring honey harvest, there is of course a massive population of bees which have to be contained and needing food. It is quite common for a single national brood chamber to be completely full of brood at that time. No stores at all. A few days of inclement weather and the massive colony will be starved. There is less chance of this happening with the 14 x 12.

External comparison of National and 14x12 (14x12 on left of photo).

Necessary space for modern prolific Queens.

So, I am not surprised that the 14 x 12 is gaining in popularity. It is a progressive development matching the improving productivity of modern queens and the increasing skills of beekeepers.

What about the often-published criticisms of the 14x12?

Brood chamber too heavy. Well certainly it's heavier of course but it has the strong handholds of the modified National. On the occasional necessary lifting, say for a Pagden or Demarée, contact a bee buddy for help or just stick with the inadequacies of the basic National. Commercial, Langstroth, Dadant, and the surprisingly frequent Long Box of 13 National frames are just as or even more cumbersome and they don't have the strong hand holds.

Single frames difficult and awkward to lift. The frames have the helpful strong and long lugs. The really heavy ones are the honey-loaded outside frames which don't have to be examined very often but as the photo shows frames are easily held and lifted by an eleven year old.

Too large for non-prolific bees. Who wants non-prolific bees? But this depressing view is seriously advocated by some with the odd opinion that they perform best in the U.K. and Ireland. Prolific bees can do well in a single national brood chamber, but a good level of beekeeping skill and experience is needed. (See photo). 14x12 makes valuable, prolific bees easier to manage.

Difficult to place in an extractor. Well using a 14x12 as a super even just to draw out comb is not a bright idea. However, I once did that with a strong hive on oil seed rape. I extracted them tangentially and had some fine drawn comb to use. But it's not a convenient routine admittedly.

Combs fall out of frames. If they do and are wired, then that's bad handling but remember the helpful use of a drawing pin described earlier. We all should know the correct way to turn over a frame whatever its weight.

12

Warré Hive

Dr Ingo Scholler

*D*r *Ingo Scholler is a hobby, city bee-keeper in Wales, UK. He keeps bees in a sustainable way and believes beekeeping promotes lifelong learning. He is a volunteer and teacher with the international bee charity, Bees for Development based in Monmouth, UK. He has between 6-8 colonies in Warré, Topbar and National hives. He has almost 6 years of beekeeping experience.*

The Warré or "People's" hive was designed by French Abbé, Emile Warré (1867 – 1951). He experimented with over 300 hive designs before settling on this format. His aim was to create a hive in which anyone could keep bees in an economical, productive and simple manner. He incorporated observations from tree colonies into the dimensions and management of his hive. In 2007, David and Patricia Heaf translated Warré's book, *"L' Apiculture Pour Tous"* into English. Further editions in Spanish, German and other languages have increased the worldwide popularity and use of the Warré hive, which in Australia is being used on a viable commercial basis.

Hive design

This is a vertical top bar hive (fig 1). The components consist of a hollow, ventilated flat or gabled roof (fig 2), covering a quilt box (fig 3), which sits on a coarse cover cloth, over the upper bee box (fig 4). The quilt enables ventilation, moisture release, and is filled with sawdust, straw or other natural insulation, also preserving the hive atmosphere and nest scent. The quilt is 100 mm deep and of the same width as the hive boxes (fig 5). These are 210 mm deep, square in shape with an internal width of 300 mm. Warré chose these dimensions as he believed they conformed to his measurements of an average bee tree cavity. Each box (fig 6) has 2 wooden handles, 2 internal rebates or battens that support 8 topbars (each 9 x 24 x 315 mm). Warré used pins or nails to fix the topbars in place. Traditionally, no frames are used, however some beekeepers have made side bars or full frames to comply with their local beekeeping regulations that may require frame removal for disease inspection. The hive boxes are all uniform, there is no larger brood box. The thickness of the box walls is no less than 20 mm, thicker is recommended in colder climates. One can use Cedar or other suitable wood types. Warré used at least 3 boxes in established colonies, however up to 5 or more can be utilised in areas with very strong nectar flows. A solid bottom board (20 mm thick or more) is used as the hive floor, with a sloped entrance (120 mm wide and 40 mm deep), and an alighting board. The whole hive is placed on a stand with legs, at least 100 mm off the ground. One can incorporate glass or perspex windows with wooden covers for observing the colony's growth and activities. Ventilation is achieved via the quilt and hollow roof. As there is no crown board, air passes through the cover cloth into the quilt and roof; the bees can add or remove propolis to control this. The hive entrance is smaller than many other hive types, allowing a slow ingress of fresh air.

Basic Management

Warré recommended inspecting the colony only once or twice per year. Useful observations regarding colony activity and health can be made at the hive entrance. A new colony is placed in 2 boxes. Bees from a swarm, split, nucleus colony, another Warré hive, or a package can be introduced. There are more complicated methods of transferring bees from other hive designs, and this information can be found in Dr David Heaf's book or website (see below). Natural foundationless new comb is drawn on to the underside of the topbars. There are various methods of guiding this comb building. A lengthwise groove can be sawn into the middle of the underside of each topbar and filled with wax, or a narrow wax starter strip inserted (fig 7). Round or triangular wooden guides, coated with wax may be nailed or glued as an alternative. When the upper box is fully drawn out, building in the next box below is commenced. No queen excluder is used and the brood nest expands downwards as honey stores are deposited above.

When the upper box is almost fully drawn, a new box is placed below: this is called nadiring or undersupering. One can assess growth by looking into the lower box with the aid of a mirror, or taking photos using a camera or mobile phone. A hive lift will allow a more comprehensive inspection from below. Fresh comb is built to the colonies own design, and as there are no side or bottom bars, it can become attached to side walls or the topbars of the box below. The latter can be reduced by sawing a further rebate into 2 sides on the upper, inner edge of each box. The combs viewed from below are of varying shapes and sizes.

The colony thus expands downwards, similar to a tree colony. At harvest time, the bees are removed from the top box by a clearer board or smoked down and this box is then removed. Traditionally one takes a whole, ripe box of honey in the autumn or spring, however unless you are in an area with very strong and persistent nectar flow, usually only one harvest is taken, provided there are enough stores for the colony to survive the coming winter. One can however remove individual combs with a special long harvesting tool by carefully cutting down each side wall to free possible attachments. In spring the floor should be cleaned or replaced with a new one. Warré used "cold-way" comb orientation in summer and "warm-way" in winter. This will vary according to the local climate and beekeepers' personal preference. One can reduce the entrance in winter and use a mouse guard; wasp guards can also be attached when required. It is useful to apply straps to avoid the hive being blown over in winter or during stormy weather, especially if the hive is situated in an exposed location.

Feeding, stores and treatments:

Warré recommended feeding disease-free honey to his bees, but only when necessary. This can be done via an inverted plastic contact, glass jam-jar or Ashforth type feeder, inserted above the upper box, by exposing some of the top cover cloth. It is easiest to use an appropriate seized eke or empty box around the feeder and then reassemble the quilt and roof; which should reduce the risk of robbing. Fondant can used in the autumn. By weighing the hive one can determine the amount of stores present and whether a harvest can be taken. A minimum of 12 kg honey stores are required to overwinter, more in colder climates. Hive lifts, some with built in scales, or a long wood clamp with a spring-type weighing scale can be used to assess the hive weight and is more accurate than hefting alone. By weighing an empty box with all topbars and adding 4 kg for wax and 1 kg for bees, one can work out the box weight without honey. The maximum weight of each topbar with honey is around 2 kg, so a total maximum of 16 kg honey per box. One can thus calculate stores by weighing the whole hive, without roof and quilt. The colony should thus be able to overwinter on 2 full boxes. Small late swarms however have survived winter on less than 1 full box under favourable conditions.

Treatments can be used in a Warré colony. Compounds requiring vaporisation can be applied from below, however the distribution may not be the same as in a conventional framed hive due to the shape of the combs. Other substances applied to topbars will require splitting boxes or insertion into or on the upper box. A mesh floor and *Varroa* monitoring board can also be used, and can be made or purchased from a supplier.

Why I like the Warré Hive

This hive can be made with simple hand tools, requires only 1 size of box and be produced at relatively low cost. One can increase the box wall thickness to suit the local conditions. Minimal intervention and disturbance allows for a more apicentric form of beekeeping and in many ways resembles the life of a colony in a tree hollow. The quilt ensures good ventilation, heat retention, moisture regulation and nest scent preservation. Free, natural comb building and harvesting, minimises the accumulation of toxins and disease infestations. At harvest the oldest comb is removed, ensuring better hive hygiene. In winter, the bees are able to move upwards in a relatively small hive, utilising the rising heat to keep warm and eat stored honey. Less honey per season is consumed compared to larger hives. If greater honey harvests are required, more Warré hives should be deployed. Minimal intervention allows the colony to control their internal environment, manage diseases by natural selection and swarming. The Warré is a well-established and successful hive recommended for both the beginner and more experienced beekeeper seeking a more bee-focused management style.

Some Pros and Cons of the Warré

Advantages	Disadvantages
Resembles tree cavity / more apicentric	Intensive conventional management difficult
Minimal intervention, 1-2 inspections per year	Cannot inspect individual frames easily
Foundationless, natural comb building / renewal	No frames, cannot reuse comb easily
Uniform, lighter boxes	Smaller honey harvests
Colony requires less stores / feeding	May require hive lift to nadir large hives
No queen excluder used	Cannot move frames between hives
Easy to build and inexpensive	Cannot extract via centrifuge easily
Swarming allows brood break	Cannot accurately predict swarming

Reading List

1. *Beekeeping for All*: Abbe Warré - English translation by David and Patricia Heaf (Northern Bee Books, 2015)

2. *Natural Beekeeping with the Warré Hive*: David Heaf (Northern Bee Books, 2013)

Internet resources

Technical drawings of an authentic Warré hive: *http://warré.biobees.com/plans.htm*

David Heaf's Warré resources: *https://warré.biobees.com/methods.htm*

The author would like to thank Paul Honigmann for his feedback on this chapter and Dr David Heaf for his ongoing support via his Yahoo Warré Forum group and Warré Beekeeping Course.

Fig 1: Working complete painted Warré hive with 2 boxes and stand of author's own design.

Fig 2: Painted gabled roof.

Fig 3: Quilt box with straw insulation.

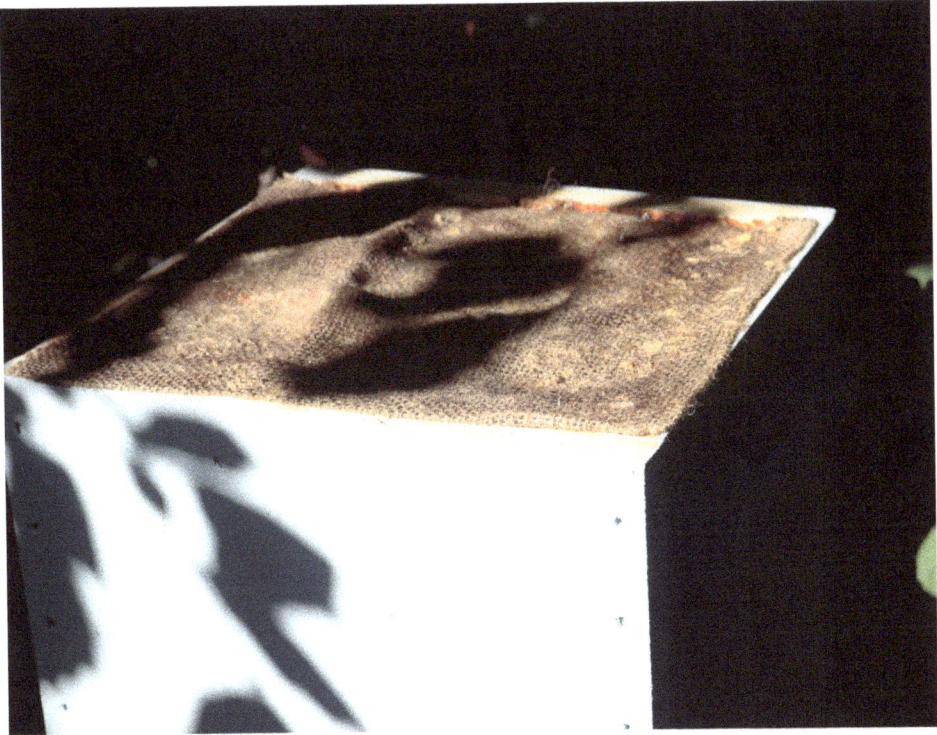

Fig 4: Top box hessian cover cloth with propolis.

Fig 5: Warré hive box showing handle and rebate.

Fig 6: Topbars with Warrét box.

Fig 7: Underside of topbars with melted wax starter strip.

13

Top Bar Hives

Chris Slade

Chris Slade has been keeping bees for over 40 years, mostly in Dorset but also with sites in Somerset and Cornwall. Currently he has about 14 hives scattered over a dozen apiaries being aware that, apart from the beekeeper, the worst enemy of a colony of bees is the one next door! Most of his hives are still Nationals but he is gradually increasing TBHs. At the moment he is experimentally joining 2 National hives to make a long hive, which might share some of the advantages of the TBH. Not having a sweet tooth, honey is a sticky embarrassment which helps cover his travelling costs. The main reason he keeps bees is that opening a hive is like meditation: everything else leaves the mind!

What? Why? Who? When? Where? How?

What is a top bar hive? It's a hive which has no frames but top bars that go where the top bars of frames would go but they abut so there's no gap between them for bees to get through. It's set on a stand so that the bars are about waist high. There are various designs, but I shall concentrate on my own which, obviously, I think is the best.

It's a horizontal half-cylinder, the top bars of which form the diameter. The length of the hive is about 3 feet and is not critical. The width is determined by the top bars, which are 17" long for compatibility with National equipment in case I want to take out some bars to make a nucleus. By applying Pi, I discovered that the comb size is almost exactly that of National brood combs.

The semi-circular U shape enables the bees to draw their comb in the catenary shape that they naturally prefer, and they are less inclined to attach the comb to the side walls of the hive than they are with the trapezoidal design which is available commercially.

The width of the top bars is about one and three eighths inches. There are various ways of encouraging the bees to draw their comb along them, although by nature they're inclined to build their comb in a curve which, like corrugated iron, must be stronger and better able to withstand movement if their home in a hollow tree is blown by the wind. Generally, I use a lump of beeswax and a soldering iron to draw a line of wax along the centre of the bar. I have sometimes bevelled bars so that they form a V shape or glued a narrow strip of wood along the line. Some people put in pegs, but this must make harvesting difficult, although the combs might be less inclined to break off if handled carelessly.

The entrance of the hive is at the sunny end. Some trapezoidal hives on sale have the entrance in the middle, which is inviting isolation starvation as the bees rear their brood near the entrance and gradually work their way back through the stores to the rear. In the depth of winter, the bees in such a hive may get through half their stores and the other half will be out of reach!

Why use a top bar hive? Originally it was designed for use in East Africa, which is why it is often called the Kenya Top Bar Hive. It can be made very cheaply from local materials once the locals have been shown how. When conventional hives, such as the Langstroth, are donated they are usually the best piece of furniture in the house and far too good to put bees in!

Mine are made mostly from recycled pallet planks and so cost virtually nothing, the most expensive parts being the fence posts forming the stand.

There's no heavy lifting involved, and everything is at waist height, so you don't have to stoop, thus you're less likely to get back problems with a TBH.

The bees create their own comb with the cell sizes THEY prefer, not the 'one size fits all' foundation we provide. Some years ago, with the assistance of my son, I measured the cell size of all the worker comb in a TBH, top right, centre, bottom left on both sides of the 16 combs. The cells ranged between 5.6mm and 5.1mm with the smallest generally towards the entrance and in the centre of the comb.

Learning from this, I haven't used foundation, apart from starter strips, in my National frames for some years and notice that the hives with naturally designed comb seem to do better than those still with foundation-based comb. Observing bees at the entrance you can see a range of size of the fliers.

An important reason for using top bar hives is that the temper of the bees is much improved as they don't know you're there! With a conventional stack of boxes, you crack them apart when they've been propolised, sending a tremor like an earthquake through the hive. You lift off the crown board. The warm cosy atmosphere wafts out and fresh air, lacking the homely pheromones, is drawn in, possibly contaminated with smoke. Light pours in from above and the bees can see a gigantic figure making rapid movements, obviously attacking the hive! If you were a bee, what would you do?

Once, on Portland, where I was managing an apiary on a communal garden, I happened to be there when there was a party. The Mayor and his good lady were interested in the bees (they may have been sponsors of the project) so I took them along to see them. I opened the TBH and, working gently from the rear, was able to show them the bees working on the comb and I broke a piece of honey laden comb off as a present for the Mayoress. None of us had any protection and I had no smoke or spray. The bees ignored us.

Who should use top bar hives, apart from Africans with the encouragement of Bees for Development and Bees Abroad? Anyone who has, or wants to avoid, back problems. Anyone who wishes to work with rather than against the bees like those control freaks who insist the bees work according to the books. Anyone who wants to keep bees but is reluctant to spend a fortune on commercial equipment. YOU!

When do you start with a TBH? In the swarming season. You can bait the hive with a drop of lemon grass oil on a blob of cotton wool. That will attract the scouts who are looking for a new home, weeks before they swarm. The entrance to the TBH being much higher than that of the National, for example, next door will be more attractive to the scouts. Don't put the bait too near the entrance or the queen might not go in, sensing a rival. Once, on Portland, the bees built their brood comb on the outside of a baited National while storing honey inside. If you visit the Island, you'll see lots of notices and car stickers advising: Keep Portland weird!

If you obtain a swarm from elsewhere you can simply remove a few bars, drop the swarm in and replace the bars. If you consider they need feeding you can cut the side off a plastic milk container, fill it with fir cones to provide rafts so the bees don't drown, place it in the rear of the hive, add syrup and make a mental note to remove the feeder before the comb building gets that far back.

When do you harvest your crop? Apart from as outlined above, I don't normally feed my bees, so I take my share of the honey when the willows and dandelions are in flower, knowing that it is truly surplus to their requirements and not partially recycled Tate & Lyle. It's easy to take the honey as it's stored at the back of the hive so you go gently forward, dropping the ripe comb into a plastic bag and slicing it from the top bar with a knife or hive tool, leaving a 'footprint' of comb to guide them in rebuilding. At that time of year most people will have sold all their honey, so you'll have the marketplace to yourself. You should get quite a bit of cut comb honey which sells for twice the price of the bottled.

Where is a good place for top bar hives? They aren't very portable so try to avoid areas of industrial agriculture, instead seek areas, preferably organic, where there will be a long and varied flow of honey and pollen. They're good for areas with badgers as those pests can't reach up and demolish a TBH as they can a National. They're good on slopes or hillsides where you have difficulty in keeping a National vertical, also where there's a flood risk. I keep a TBH at Ourganics, a permaculture holding on former water meadows. The owner, Pat, also keeps bees there. On one occasion there was a flood and her National was swept away and the bees drowned while my TBH, a few yards away, was above water level.

How do you cope with *Varroa*? The bees in top bar hives seem not to be afflicted as noticeably as in conventional hives, but I don't know whether this is in fact so. In order to monitor mite drop I built a cubic TBH like a coffin with a mesh floor within curved to the catenary shape as described and with a tray below which could be withdrawn, and the mites counted. I placed it at the Bee Happy Plants nursery with the intention that my apprentice, Sarah, the proprietor, should make regular inspections and record the mite drop. Unfortunately, she is more interested in pollen than in mites and was distracted by the loads that fell through which she gathered and spread on a diary so she could see what was coming in when. So, to answer the question at the beginning of this paragraph, I chuck in a tea bag of thymol crystals, as invented by Ron Brown (This is not a licensed treatment), in the Autumn and, if I feel the need, dribble oxalic between the occupied bars around Christmas.

How to get a crop of propolis: remove one bar and ease the remainder apart by a smidgeon to take up the space. On your next visit you can scrape the sides of the bars with your hive tool and harvest a matchbox or two of propolis. Sometimes, usually during the wasp season, the bees will use propolis to reduce the entrance to make it more defensible as can be seen in the next picture. They remove the propolis when no longer needed.

How to deal with off-line comb? If you find it early enough you can ease it on-line with a hive tool and the bees will have secured it by your next visit. Alternatively, you could cut off the curved bit and place that bar between two combs that are aligned as you would wish. Occasionally, if you've not inspected for a long time, you'll find cross-combing. You need to lift out several bars at once, remove the comb and secure it on-line with string or elastic bands, alternating such combs with good ones in the hive.

How to reduce swarming. Make sure they have plenty of space. Build the hive with a corked hole at the rear that can be opened and used as an entrance if the hive is split with a few bars with a queen cell at that end and a division board inserted. Alternatively take out a nucleus and hive it in a National nuc box (trimming the edges of the comb).

How to renew comb. When you take your harvest in the Spring, place one or two of the sliced off bars at the front of the hive which the bees can draw as they move forward. At that time the brood nest will still be relatively small, and any unused dark combs can be sliced off, leaving a footprint, and replaced.

How to do the woodwork. The most difficult bit is getting the planks off the pallets by trial and error. You need to do some arithmetic or draw a diagram and apply a protractor to work out how many planks you need, and to which angle you should cut or plane them. They don't have to butt perfectly as you can apply beeswax and a hot iron to the outside to preserve the woodwork and fill any gaps. The bees will attend to roughness and irregularities on the inside by applying propolis. The planks need to be attached to fairly solid ends made of overlapped pallet planks or thick ply. The ends should be the thickness of a top bar higher than the sides in order to hold the bars in place.

Don't forget to make an entrance! You can experiment with this. Normally I use a circle about the size of finger to thumb (I don't know what that is in metric) but occasionally during the wasp season the bees will make a screen of propolis to reduce it to little more than a single beeway. On one TBH I applied propolis with a hot iron to make a smiley face with the curved mouth forming the entrance. On another hive, where I thought there was danger of robbery or wasps, I made a Fibonacci spiral of nine holes into which you could just fit a little finger. The bees loved it!

As you will see in the photos on the next page, in a recent hive I made a row of small, defensible, entrance holes with fishing line stretched beneath in case they wish to use it to get rid of mites as suggested by Stuart Roweth who designed the BeeGym.

How to support it? The first one I made, 20 years ago, was sat on plastic binder twine slung between fence posts. After about a year the twine broke! The next one, for the same hive, was a solid wooden stand that still looks good but was a little difficult to make. More recently I've been using fence posts either placed vertically with holes drilled and stout wire through them to support the hive or else crossed to make an X with another, short, post at a right angle to ensure it doesn't lean. Bear in mind that the hive must be horizontal as the bees will draw their comb vertically. Also bear in mind that as there is no frame or wiring you must keep the comb vertical at all times as it's likely to break off if you don't, particularly if it's heavily laden with brood or honey.

So, as you will have read, there's no hard and fast design and method for making and using top bar hives. Just do it your way, bearing in mind that it's for your enjoyment and that of the bees.

14

Dartington Long Deep Hive

Robin Dartington

1.0 Evolution of the Long-Deep hive form

After 10 years in National hives, in 1975 my bees moved to a 5-storey town house with a flat roof reached by ladder. Everything had to be kept on the roof in the open, as going down five storeys and back up with an extra super was not much fun. But empty boxes blew off in a high wind!

I changed to 'deep brood' from 'brood and a half' and screwed an empty deep box to the back of each occupied hive, ready for artificial swarming – so creating a long box with an entrance at each end. The fixed central division became irksome, so I cut it away, then made my first Long Deep hives from recovered floorboards[1], inserting a division board of thin plywood that could be placed wherever wanted. I made half-sized 'honeyboxes' taking only 6 shallow frames that were easy to hang from one hand when on the ladder. All boxes were 'top-bee space' so the half-sized 'cover boards' could be simple pieces of 18mm ply.

---

1 | *They are still used after 40 years for storing frames!*

There is nothing new in beekeeping, so I searched for 'long-deep hives' in the literature. Nearest was Cheshire's hive of 1886 which held 8 brood frames plus a 3-frame nucleus behind a moveable divider as the strain of bees then only made small nests – and Abbott's Combination hive of 1878, so called because it stored honey both 'behind and above' the brood. There were thousands of Layens hives in France and Spain, which had a broad (rather than long) body with an entrance in the middle and no supers. Russian long hives held 20 to 24 frames and had supers – but the translation of the only Russian book I could find was very difficult. I concluded I had to devise a management system for 'long-deep' hives that could hold up to 21 deep frames between two insulating dummies and be covered by a row of half-sized 'honeyboxes'.

2.0 Suitability of the Dartington Long Deep (DLD) hive for modern beekeeping

Beekeeping - and in particular beekeepers - have changed over the last 30 years. Growing affluence means that fewer beekeepers look for a side-line income from bees - more are attracted to the fascination of exploring this ancient and complex life-form as a way to contact nature. The advantages of a Long Deep hive are relevant to these new beekeepers - they are not so physically strong, they value neatness more, they do not want to move their hives around to get a bigger honey crop.

The hive body is equal to two Deep National hives pushed together, allowing a colony as much space for brood as it wants, and avoiding need for queen excluders provided there are sufficient drone cells in the brood frames, otherwise the queen lays drone eggs in the honeyboxes. The long box makes it easy to divide the colony to control swarming for swarm control or to raise a spare queen in a rear nucleus.

The overall length of a DLD hive is only twice its height when using one layer of honeyboxes. The length of a true 'long hive' using standard frames and no supers is six times the height. It is a common mistake to confuse the DLD hive with a simple 'long hive' – these two hive-forms have nothing in common.

This hive is **convenient** as it avoids need for a second empty body for swarm control – division into 'swarm' and 'parent' involves moving only single frames sideways, one at a time, before inserting a division board. It is **economic** as one long-deep hive is cheaper than two separate hives – and if carefully managed, can produce a large honey harvest in a good year. The hive is **safe** as the heaviest lift is a full honeybox weighing only 8kgms.

This form of hive is suitable for **garden beekeeping** as you can keep everything needed through the season under the two-part roof, avoiding untidiness. Being double-length, it does not become excessively high' with supers, so the body is raised on fixed legs to a convenient working height, avoiding back strain. The DLD is a static hive, not designed for moving around to different crops[2].

---

2 | *Beekeepers who migrate hives would find a three-quarters long box on a loose stand, taking 15 frames and three honeyboxes, easier to lift between two people - the Dartington Migratory Deep hive.*

WINTER

A. Beekeeper 'prunes' the nest back to immature state of 9 frames

B. Beekeeper adds foundation frames

C. bees build new combs to restore the nest to mature size of 15 frames of deep comb

E2. 'parent' rears a NEW QUEEN

D. Beekeeper divides the nest into 'swarm' and 'parent'

G. Reunited colony provisions the nest for winter

E1. 'Swarm' develops NEW BROOD NEST in new combs

F. Beekeeper re-unites 'swarm' and 'parent' with NEW QUEEN in NEW BROOD NEST

SUMMER

THE AIM OF NEW BEEKEEPING - THE 'MANAGED' COLONY CYCLE

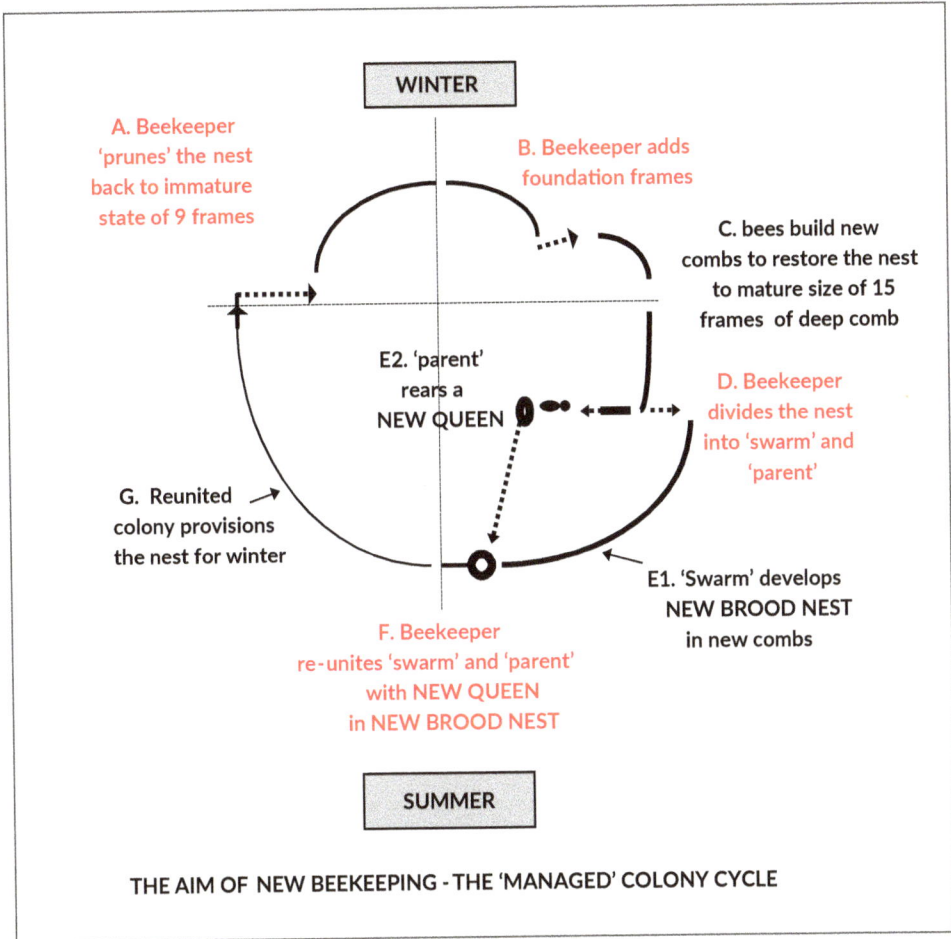

A Long-Deep hive is very suitable for family beekeeping, as even a full honeybox can be lifted by children – and by the elderly, so no need for Grandpa to give up.

3.0 Summary advantages and disadvantages of a long-deep hive

Advantages

A 'long-deep hive' makes operating easy:

- The attached legs raise the body to waist height, avoiding bending.

- The heaviest weight to be lifted is a half size 'honeybox' (8 kgs).

- The deep frames allow the bees to create a one-piece spherical brood nest, the most efficient shape for conserving heat.

- The bees are able to pack honey at both sides and above the brood nest, so helping to maintain a constant temperature in the brood nest.

- Containing the block of deep frames between two insulating dummy frames allows flexibility to suit larger and smaller colonies – no need for 'one size fits all'.

- Only a single deep frame at a time needs to be lifted or slid sideways when separating 'the queen' from 'the brood' for swarm control.

- Swarm and parent are completely separated by merely inserting the division board, without need for a second complete hive or nucleus box.

- Re-uniting swarm and parent involves only removing the divider as that joins the backs of the two colonies and the guard bees at the front are unaffected.

- An occupied honeybox holding 5 Manley frames or, 6 shallow Hoffman frames - becomes a 'queen mating box' by merely inserting a 'clearer board' (with the hole covered over) as a floor, and fitting a disc entrance in one wall or nicking the rim of the clearer to create an entrance.

- The two 'carry-boxes' enable efficient winter storage of the 12 deep frames that are added and removed each season – and can act temporarily as deep nucleus boxes in summer - or for holding deep frames containing honey for clearing out by 'controlled robbing'.

- The hive is designed for making at home by simply assembling pieces of plywood and attaching softwood battens, without need to form any joints[3]. With a little more trouble, it can be made wholly of softwood.

- Unprotected plywood and softwood are not durable. All parts of a DLD hive should be painted externally with 'garden shades' - and preferably internally with floor paint.

Disadvantages

A long-deep hive is unusual:

Beekeeping tutors only teach conventional beekeeping, using National hives – local advice and support for 'New Beekeeping' may be difficult to find;

A Dartington Long Deep hive is only sold as a complete beekeeping kit, making the first buy expensive;

The hive was designed to be made at home – but this requires time and tools such as long window clamps;

The DLD hive is too heavy to be convenient for moving to crops – the three-quarters length Dartington Migratory Hive is recommended.

Extracting deep frames requires access to a large enough extractor, such as a 9-frame radial fitted with loose tangential screens.

---

3 | *A detailed 60-page booklet, Construction Information for Dartington Hives, is published by Northern Bee Books.*

HoneyWorks Bee Training Centre, Hitchin Herts.

4.0 Design features of a Dartington Long Deep hive.

4.1 Materials.

The hive is designed to be made at home from 18mm and 9mm and 6 mm plywood, plus attached softwood sections available from Wickes and other building suppliers. The booklet, 'Construction Information for Dartington Hives'[4], shows how to cut the pieces economically from full plywood sheets which merchants often have a wall mounted saw to cut sheets to convenient sizes for transport – or even to cut right down to a full set of finished pieces if you ask nicely.

Plywood is used in order to get large stable pieces, but the body can be made completely from softwood as that give's better insulation. (I am now fitting 25mm foil-faced building insulation permanently inside the front and back walls – and contemplating fitting insulation on the outside of the long side walls, covered by painted thin ply or softwood to improve appearance).

Hives should be fully painted externally with 'garden shades'.

Beekeepers who are not up for DIY may buy hives fully assembled (but unpainted) from a bee supplier. The price may seem high but, as the DLD hive is the tool for New Beekeeping,

---

it is sold only as a complete kit for managing a colony through the whole year - whereas the beginner who buys a National has to buy a second one (or at least a nucleus box) to control swarming in subsequent years.

4.2 Special features of a Dartington hive

Anyone can make a double length box to take deep frames and so have a long-deep hive[5]. But some special features, worked out over 40 years, are needed before it can be called a DLD hive.

- **Roofs** that overhang the sides ('eaves), so avoiding rain tricking down the hive sides. Insulation fitted permanently under the roof-top;

- Four '**honeyboxes**', each 475m wide x 233mm long to hold 5 Manley shallow frames or 6 Hoffman shallows, with space to loosen the first frame;

- Solid 18mm thick '**cover boards**' 471mm wide x 230mm long, each covering a honeybox without risk of over-sailing;

- Two '**clearer boards**', 6mm ply rimmed with 12mm battens to create top bee space, 471mm wide x 230mm long, with hole for a Porter bee escape.

- A '**top-bee space hive body**' 475mm wide x 950mm long, end and side walls 18mm thick; tops of the side walls widened to 52mm by attached 34mm battens whose tops are rebated to receive the lugs of deep frames whilst leaving the top edges of the side walls 20mm wide, so avoiding risk of damage from a hive tool when levering up honeyboxes.

- 2 deep '**dummy frames**', 35mm thick overall, including Hoffman edges, to contain and insulate the ends of the block of occupied frames.

- A 44mm high, full width '**entrance tunnel**' at each end of the body to provide weatherproof landing, leading to an 8mm wide slot in the hive floor that keeps the hive permanently mouse proof.

- A detachable '**landing board**' with a 44mm high attached facia that closes the tunnel entrance when the landing board is inserted upside down.

- A '**solid floor**' to the body up to the entrance slot at each end, to provide lateral stiffness to the body and a mesh floor between the entrance slots to provide permanent ventilation.

- A 9mm thick, 569mm square, '**division board**' that can also act as a *Varroa* **board** when slid in horizontally behind the landing board.

- Attached '**legs**' or separate stand that lifts the body to waist height to avoid the beekeeper bending during operations.

- A 5th **honeybox** and 5th **cover board** as spares or to act as a mating hive to rear a spare queen.

---

5 | *Bottom bee space Long National hives are available from cornishhoney.co.uk*

DLD divided - 'swarm' to the left, 'parent' to the right, ready for insertion of the Division Board in the central gap.

- Two '**carryboxes**', 475x 233mm long, each to winter 6 deep frames in readiness for adding to the hive during the season –can act as 'nucleus hives' when fitted with disc entrance and covered by a spare cover board.

- A queen introduction cage.

- A drone fork.

- Four circular plastic nucleus entrances – drill 18mm holes in the end walls of two honeyboxes and of the two carry boxes and fit the plastic entrances for use as mating hives and nucleus hives.

- Two framed National wire queen excluders –to be absolutely sure no brood gets into honeyboxes – not unframed plastic as the proportion of hole to solid will impede air movements.

15

Rose Hive Method

Julie Elkin

I began beekeeping with North Devon beekeepers in 1974,. I started with a National and a WBC but soon decided on the Nationals, WBC was too cumbersome for me and Nationals are easier to put into the back of an estate car. I have kept bees in the West Country. After a work dictated break from beekeeping during the 90s, I resumed beekeeping in 1999 and found Varroa had arrived and changed both bees and beekeepers. For me, excessive reliance on chemical treatments wasn't the right way to keep bees so I started looking at alternative, more bee friendly ways. When The Rose Hive Method Book was published in 2010 I knew my experiments were in line with Tim Rowe's aims and method and have been happily keeping bees in Rose hives since then and our branch apiary includes a Rose hive which is proving popular.

The Rose hive is not just another box of slightly varying proportions in which to keep bees; it is the 'method' that requires a different way of looking at bees, responding to their needs with the aim of reducing stress and enabling them to stay healthy and productive. The Rose hive facilitates this philosophy.

It all began when Tim Rowe, a commercial beekeeper with around 100 colonies in County Cork. Ireland became increasingly concerned about the numerous problems bees were facing worldwide. He decided that we all needed to take a fresh look at how we had been taught and how we continued to keep our bees. He felt that although individual beekeepers may be unable to address the wider environmental issues damaging pollinating insects we could certainly improve the way in which we kept our bees. He started by studying the way in which bees have lived and survived in the wild for millions of years and concluded that conventional beekeeping was suppressing many of the bees natural instincts, stressing them and making them more prone to disease. He abandoned the use of queen excluders and started experimenting with different sized boxes. The results of all this experience were published in 2010 in his book 'The Rose Hive Method: Challenging Conventional Beekeeping.' He does not claim to have the definitive answers to all honeybee problems and accepts that his ideas may not suit all bees and beekeepers but it has worked for him. When a fellow beekeeper gave me the book shortly after publication it was a revelation, a truly 'Eureka' moment. I promptly ordered my first Rose boxes and frames and I now, as a hobbyist beekeeper, use only Rose hives.

I began beekeeping in 1974 with Nationals in the conventional way and kept bees happily with few problems until 1992 when changing circumstances led to a break in my beekeeping. Resuming in 1999 I was immediately aware that the beekeeping world was very different to the one I had left seven years previously; *Varroa* had caused huge problems but I found it hard to accept many of the practices I was being urged to adopt (e.g. drone culling, excessive chemical treatments etc) to combat the problems the bees were facing. Like Tim I was looking for a more bee friendly way to keep my bees so I went back to studying and reading widely. I also abandoned the use of queen excluders, dabbled with Top bar hives and then came Tim's book with all the answers that I was seeking.

So what is a Rose hive? It is a very simple hive with boxes all the same size, an OSB, one size box system which is the key to the method. No queen excluders are used and there is no distinction between brood and super boxes, they are all the same. The bees choose where they want to start the brood nest and how and where they want to expand it as they would do in the wild.

Natural shape of the broodnest

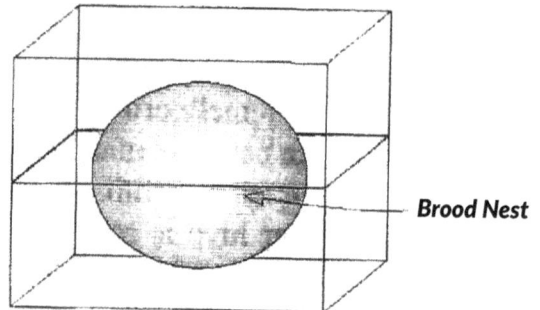

Brood Nest

Two Rose boxes are equivalent to a National brood and a half with 12 Hoffman frames in each box. Top and bottom bars are the same as National Hoffman frames but have shorter sidebars for the Rose.

Rose boxes fit with National floors, crownboards and roofs. Tim recommends that 25% of the frames have wax starter strips instead of full foundation. This gives the bees the opportunity to draw out easily as much drone brood as they want; this differs from conventional methods which restricts the production of drones. He is convinced that bees are happier and less stressed if they can choose how many drones they need. I also believe this to be true and it has the additional benefit that more drones leads to more successful mating. For a beekeeper looking for a 'middle road' between conventional and 'natural' methods of beekeeping the Rose Hive Method as Tim has outlined it offers a happy compromise for both bees and beekeeper.

The Rose Hive Year

A Rose colony usually winters on 2 boxes and a very large colony maybe on 3. Ideally the bees winter on their own natural stores as Rose hive beekeepers believe that the bees' health is undermined by the excessive feeding of white sugar. As always with beekeeping there

Comparison of the size of Rose Hive and Modified National. Rose with two boxes on the right and National brood and a half on the left.

View across the top of the 12 Hoffman frames with the expanding brood nest in early spring.

may be times of prolonged bad weather and lack of forage that may necessitate feeding colonies. At the spring clean the boxes are reversed and/or the frames are re-arranged so that the older combs are in the upper box. Instead of using disruptive techniques such as Bailey comb change or Shook swarm for comb renewal the older comb is worked to the top of the hive and then removed for extraction and rendering ensuring that no combs are ever more than 3 years old. Working up through the hive means that the oldest box of comb is now getting higher in the stack ready for removal at the end of the season. Also at this time I take the opportunity to remove any very scruffy combs including the free comb as this is unsuitable for extracting. The spring clean is the only intrusive and stressful manipulation in the year of a Rose colony.

Up until mid June, ie. the peak of the brood nest expansion, new boxes of foundation are put underneath or into the middle of the brood nest , spreading the nest vertically. This is hard to do at first when you have been brought up on the rule 'never split the brood nest' but it works. The bees do not object and very rapidly draw out the foundation. After mid June as the brood nest reaches its maximum size and starts to shrink additional boxes of foundation and/or drawn comb are placed above the brood nest for honey storage.

The advantages of using the Rose method

- The simplicity of an OSB system with all parts being interchangeable;

- Without a queen excluder there is no congestion in the brood nest, the freedom to expand the nest as and where the bees choose reduces stress and the queen is no more difficult to find than in a colony on a brood and a half;

- Less congestion can reduce the tendency to swarm;

- Placing a new box into the middle of the nest saves the workers having to move the dome of stores surrounding the top of the brood to expand the brood area;

- Colonies build up rapidly as the new boxes are given where and when the bees need them;

- Making splits for nuc rearing or swarm control is very straightforward when all the frames and boxes are the same size;

Diagram adding a new box in the centre of the brood nest

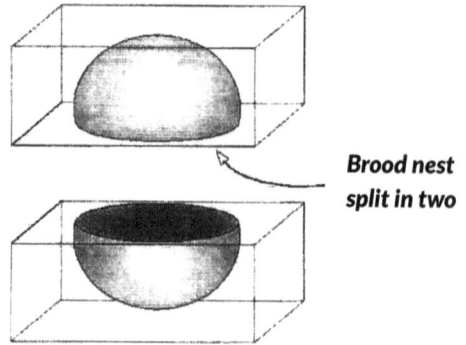

Brood nest split in two

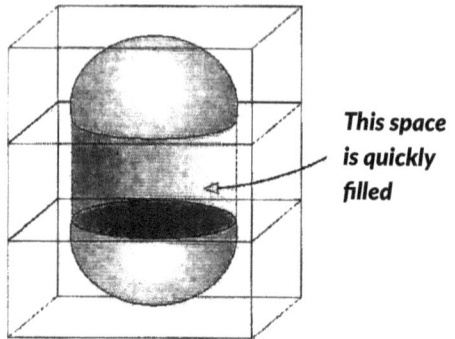

This space is quickly filled

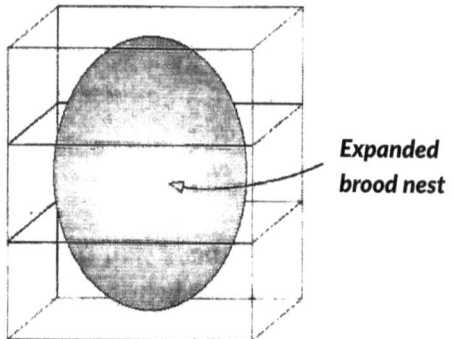

Expanded brood nest

- It is very easy to convert to Rose from National or Commercial hives as floors, crown-boards and roofs are all compatible. The Rose method stresses the need for crown-boards to be made of thin porous wood to allow water vapour to pass through easily and without holes to avoid through draughts and;

- The boxes are simple to construct and Tim Rowe's book contains full plans and instructions making the Rose a very economical hive. They are also cheaper to buy than most other boxes.

The disadvantages of using the Rose Method

- The boxes although smaller than a national brood box are heavier. A Rose box of capped honey weighs around 25kg (50lbs) so perhaps not the ideal choice for an ageing bee-keeper but I can still manage them!

- Marine plywood is used for 2 sides of the box and this does not wear well or offer the same insulation as cedar. The boxes do need to be protected by painting or using another bee friendly coating. My 6 year old boxes now need these sides replaced but it is a simple job and Thornes will provide replacement boards. Thornes are the sole UK supplier and they are also looking for alternative material to use to overcome this problem;

- Some beekeepers have claimed that bees do not winter well in Roses but my bees in cold and damp north Devon have had no problems;

- The lack of separation between brood and stores is a concern for some beekeepers but the stores are still above the brood and it is no more difficult removing frames for extraction or cut comb than in any other type of hive. The queen may be able to enjoy a 'walkabout' but she doesn't lay outside the immediate brood area the workers have prepared for her and so there is no danger of brood being in the honey stores and

- A swarm is the ideal way to start your first Rose colony as the majority of nucs available to buy are on national brood frames and this may be a little challenging for a beginner to convert to a Rose box. Tim's book does give good advice on the way to do this.

I have now been using Rose hives and following Tim's method for 7 years and I believe the advantages of using Rose hives heavily outweigh the disadvantages for both bees and bee-keeper, my bees have thrived with this system. If you are about to embark on a lifetime of happy beekeeping I would encourage you to read Tim Rowe's book 'The Rose Hive Method' and as he says 'allow your bees to do as bees want to do'.

16

ZEST Hive Design

Bill Summers

I was born in 1944 and am a retired Architect. I have kept bees in Dorset since 1972 and ran up to 80 colonies in B.S. Nationals until 2010. As the inventor of the ZEST hive I presently keep up to 18 colonies in ZEST hives.

I have kept bees since 1972 in traditional thin walled hives losing a third of them every winter until 2012. As an Architect before retirement I had always considered traditional hives as a design challenge waiting to be accepted when I had the time. A hive or a house is just a habitation for a biological system with environmental control needs to assist its inhabitants.

The American architect and General Systems Theorist Buckminster Fuller said that human technical engineering advance is only made by doing **MORE** with **LESS**. The process by which this is achieved is by engaging in the process of **Design/Science** in which science takes things apart to understand them and design puts things back together to make a useful object. Everything can be reduced to energy input compared to energy output as a measure of the effectiveness of the design.

The ZEST hive's name is an acronym from **Z**ero **E**nergy and **S**us**T**ainable and refers to the ambition that it will over its lifetime gather more energy as nectar and pollen than was expended in making it.

I understand that nationally we lose about a third of our colonies every winter. This should be telling us something about the hive designs that we currently use. If a traditional wood hive such as a B.S. National has a resistance (R) to the passage of heat of 1 then a poly hive has an R of 3 and a ZEST hive of 39. Not only that, but a ZEST also has a thermal capacity the same as a storage heater, assisting the bees in thermo-regulating the brood temperature. New beekeeper's mentors teach their pupils that which they were taught rather than move into a new understanding that moves into a modern (perhaps IKEA) aesthetic.

The capital cost of starting up beekeeping is beyond the financial reach of many. We pay hundreds of pounds for wood hive boxes which we then leave out in the rain. We fill them with highly engineered wood frames with wax foundation.

Costs do not end with the hive and its fittings.

There can be as much stainless steel furniture in a traditional hive beekeepers house as in the local medical centre. We all have to buy our own centrifugal extractor, because the Association's one is always being used by a Committee member just when you want to use it. The collective costs of all this extracting equipment, just so BKA's members can each spend a couple of hours a year extracting honey, can amount to a small mortgage. Once obtained, the extractors use is not without casualties as an unbalanced extractor tries to throw its user about the kitchen, breaking the crockery and indeed marriages when the sticky mess offence is also taken into account.

Cost

In considering the cost of bee hives and the running of them we can return to the doing MORE with LESS concept of Buckminster Fuller. The cost of any object that you buy represents the embodied energy in all its various forms that it took in its making and delivery.

The ZEST hive capital costs (2019) are as follows:

1. £100. <u>External envelope</u> from a Builders Merchant made from:-

    A. 1# 600x600 paving slab foundation.

    B. 4# 440x215x100 heavy concrete block supports.

    C. 2# 1200x100x50 wood beams.

    D. 24# 600x215x100 air entrained insulation blocks for floor, walls and roof.

    E. A (scrap) metal sheet roof covers the whole assembly and is held down with ropes.

2. £128. <u>Internal fittings</u> from The ZEST Hive Co. Ltd. comprising:

    A. 2# boxes of 12 patented plastic wax foundation less frames made from T-bars, the tails of which support the spine of the bee's natural honeycomb from where it is drawn. These are double brood depth, but in 3 equal parts.

They fit a B.S. National double brood, a 14x12 and a shallow and 3 shallows. (£72). *With average DIY skills all the ZEST plastic fittings can be replaced with wood or even bamboo frames held together with paper clips. The wood frames can be battens, gun stapled together with grooves to receive wax starter strips. They can sit on a DIY wood carrier frame with bee entry and ventilation slots.*

B. A box of 4 queen excluders. 2 of which are turned into partitions with a bolted on Corex board. (£36). *Queen excluders can be cut from traditional ones and put into a frame that fits tightly into the ZEST hive cross section.*

C. A Carrier frame to support the 24 frames and carry the load onto the block walls. (£20). *This can be replaced with wood battens and the bee entry and ventilation points drilled through the walls at high level and a virgin cage inserted into each, configured to allow access and or ventilation.*

This arrangement for £228 gives an internal ZEST hive volume of 5 B.S. Brood boxes (for 2 ZEST colonies if desired for overwintering) You are invited to cost out a traditional hive of such a volume and compare it... before you then consider maintaining and running them which, for a ZEST, is a fraction of others.

The ZEST is simple, stands alone and needs no maintenance, just infrequent visits to add more frames and to take the harvest.

Better by Design for the Bees

Bee health in a ZEST is improved by providing a warm and dry environment. It has top bee entry and ventilation so that no stack effect occurs internally.

The thermal insulation and mass of the external block envelope assists the bees in thermo-regulating the brood nest. From my observation, the bees draw out natural honeycomb at 4.9mm on average rather than the 5. 4mm.of wax foundation, further reducing the pupa hatch time, because the bees are smaller.

This potentially speeds up the biological process reducing the time needed for the pupa to mature and emerge. This could impact negatively on the *Varroa*'s ability to sustain their numbers. We have ZEST hives that have not been treated against *Varroa* for 5 years in which there are none or few *Varroa* in the floor debris and little or no deformed wing virus. In my experience a ZEST does not seem to sustain a *Varroa* population and any found in the hive debris may have been brought in by visiting drones.

The ZEST hive can be a "let alone" hive, and little harm is done by the stealthy inspection of a ZEST.

Individual frames can be added as needed by the bees which prevents the cold shock of adding a whole box.

My Winter colony losses in a ZEST are a small fraction of those I used to have in thin walled hives and the use of stores is at a minimum. One ZEST was taken through November, December, January, and February of 2016/2017 and used only about 4kgs of honey stores.

Being 2 full brood frames deep in a long hive configuration a compact brood nest can be obtained with a small ratio of surface area to volume. A sphere brood nest shape is the best attainable for energy conservation.

Better by Design for the Beekeeper

The ZEST can be an entirely DIY hive for those who are good at it being made from readily available, cheap materials. It is maintenance free requiring no shed time in the dark winter months. There are no wax foundation costs, I find the bees are better tempered and so slow to swarm and that that the term "swarm inspections" becomes an anachronism.

Hive equipment storage space is not needed as it is all kept in the hives.

Osteopathy bills are reduced when the maximum load lift from a ZEST is only 4Kgs, being that of a full honey frame. To gather this honey only one visit is needed. Porter bee escapes are redundant. Brushing off the bees and hurrying off with the honey before they catch on to what is happening is the plan.

The external envelope and internal components are not only maintenance free, but are robust against predators, livestock, storms and theft.

If the colony in a ZEST is strong at the end of July when the honey flow has largely finished and with time on their hands the bees can be split as an artificial swarm with the artificial swarm residue placed at the "other" end of the ZEST with a new mated and laying queen or allowed to re-queen naturally. These are then overwintered giving unaccustomed options in the spring caused by an abundance of bees.

These options for a ZEST include:

1. Uniting them again around the best queen just as the spring honey flow begins and redeploying the spare queen.

2. Keeping the two colonies at each end of the ZEST and putting traditional wood boxes and frames as supers on them with a traditional queen excluder. When we did this the *Varroa* count went up, so beware.

3. Keeping the two colonies, taking sealed honey and use them as larder hives.

4. Sell or give a colony to a stranger or friend with an empty ZEST. When ZEST colonies nearly always survive the winter they become less precious.

5. Redeploying one of the queens as the swarming season starts so the queen less colony make large supersede cells, as opposed to small swarm cells, caused by the "Cloake deception" where the close presence of the remaining queen makes them think they are only superseding.

The ZEST hive has a slightly more consistent temperature but is considerably more humid. The bees in the ZEST appear to control the humidity, because its RH rises when the temperature does, when it should fall. It does this in the traditional hive which is normal. High humidity impacts on *Varroa* numbers for reasons not known.

Buckminster Fuller said that to better the world we need to design things and systems that will do so. The world was never changed by talking to it. The planets ecosystem is a symbiotic one. We deserve our place within it to foster life, support evolution and postpone death. No species is an island, but we are the only one that knows that. We are special now. There can be a no more comprehensive and mutually beneficial symbiotic relationship on the planet than ours with honeybees. Much is made by beekeepers of the ambition to emulate the concept of "Natural" when it comes to bee habitats such as caves and holes in trees, but bees struggle to survive and thrive happily in many of them.' We can help them on this matter with the design of the ZEST hive which, with sympathetic management is conducive to both their needs and ours.

THE ZEST VARROA EXTINCTION/SURVIVAL DIAGRAM

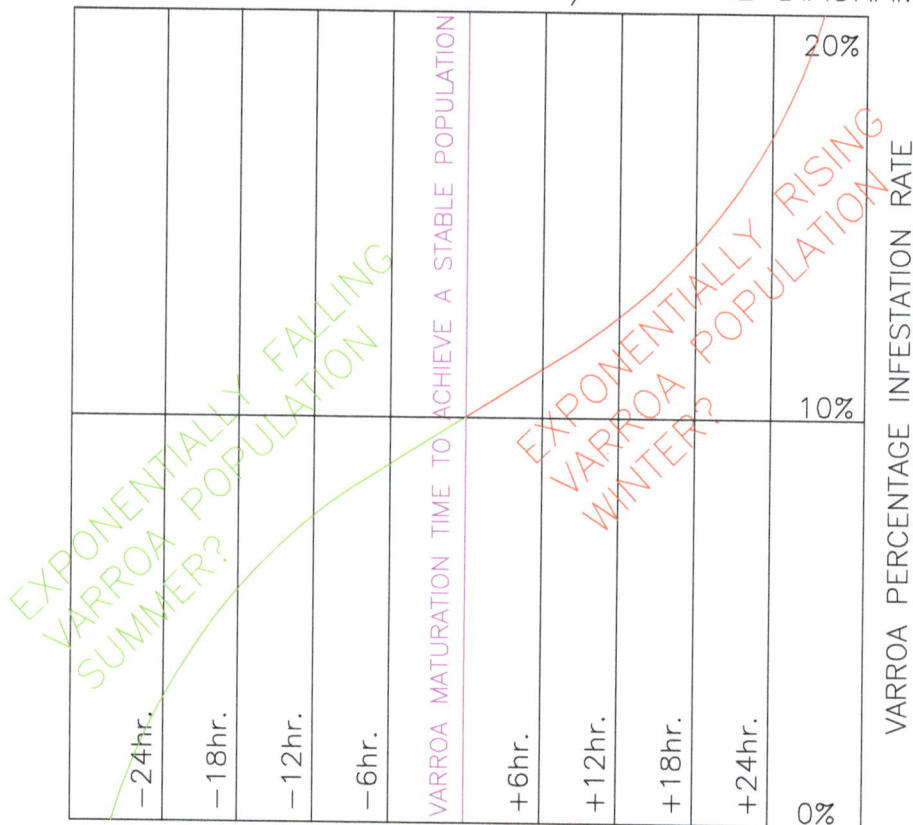

QUICKER ← BEE PUPATION TIME → SLOWER
WARMER   COOLER
SMALLER CELLS   LARGER CELLS
(HIGHER RH)   (LOWER RH)

Above 37 deg. the pupa die.
35 deg. 10–11 days as pupa.(summer?) 96–98% surviving
31 deg. 14–15 days as pupa.(winter?) 89–100% surviving.
Below 29 deg. the pupa die.
Therefore winter is varroa breeding heaven due to more pupation time for varroa to mature.
Google:– "Synaptic organisation in the adult bee brain is influenced by brood temperature control during pupal development" and view table 1.

THE ZEST HIVE BASIC DESIGN

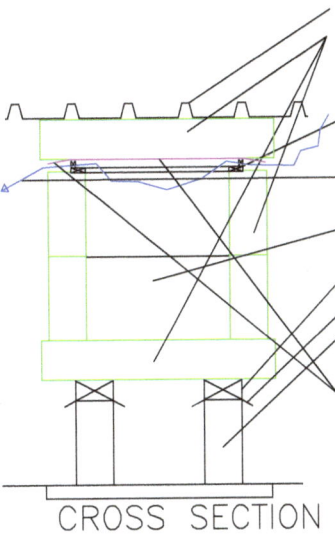

The Merits of a ZEST hive
1. Metal roof sheet from scrap.
2. Standard insulation wall blocks used as roof, floor and walls to insulate the colony and assist in thermo regulation.
3. Carrier frame with multiple top entrances for bee access and for adjustable trickle cross ventilation, housing up to four colonies. No stack effect to cool colony.
4. Void for one of three possible types of ZEST frame. i.e wood, bamboo or plastic.
5. Wood bearers on an overhanging vermin guard/damp proof course.
6. Concrete blocks on a single paving slab raising the work position.
7. Plastic cover sheet

Notes:
Maximum weight lift of 4kgs. honey in frame. Management to be 2 overwintered colonies. One with new queen and one with old.

CROSS SECTION

ALTERNATIVE FRAME TYPES
DIY STAPLED WOOD SPACER FRAMES FOR MACHINE TECHNOLOGY.
DIY COLLAPSIBLE BAMBOO GRAVITY FRAME FOR THIRD WORLD USE.
PLASTIC FRAME FROM THE ZEST HIVE CO. LTD.

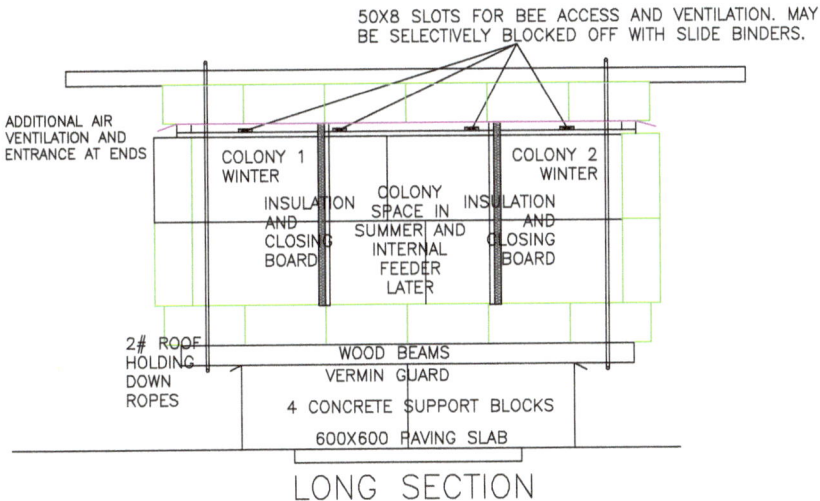

50X8 SLOTS FOR BEE ACCESS AND VENTILATION. MAY BE SELECTIVELY BLOCKED OFF WITH SLIDE BINDERS.

ADDITIONAL AIR VENTILATION AND ENTRANCE AT ENDS

COLONY 1 WINTER

INSULATION AND CLOSING BOARD

COLONY SPACE IN SUMMER AND INTERNAL FEEDER LATER

INSULATION AND CLOSING BOARD

COLONY 2 WINTER

2# ROOF HOLDING DOWN ROPES

WOOD BEAMS
VERMIN GUARD
4 CONCRETE SUPPORT BLOCKS
600X600 PAVING SLAB

LONG SECTION

AVAILABLE ZEST HIVE PARTS

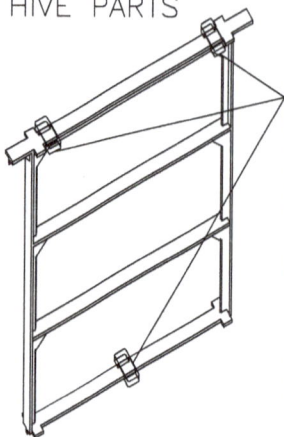

BOXES OF 12 B.S. UNIVERSAL PATENTED PLASTIC FRAMES FOR NATURAL HONEYCOMB MAKING

The standard plastic ZEST frames are at 35mm c/c. 3 snap on spacers can be added to the frames after the comb is drawn out to give 45mm. c/c for exclusive honey storage outside the brood area. These can be deployed in B.S.National hives boxes as follows:-
2 B.S. Brood boxes.
3 B.S. Supers or
1 B.S. Deep brood box with a super
or
in a full ZEST hive.
www.thezesthive.com

CUSTOMER AQUIRE IF FULL ZEST HIVE IS REQUIRED

The customer will also need to source the following.
A. 1# 600x600 paving slab.
B. 4# 440x215x100 concrete blocks.
C. 2# 1200x100x50 wood bearers.
D. 24# 600x215x100 lightweight insulating blocks.
E. 1800min. x1000 corrugated roof sheet
F. 2# roof holding down ropes
G. 2# d.p.c vermin guards on item B above.
H. 3 # 374x6 ply or plastic strips to act as feed ramp and air block under partitions.
I. 1# 1500x500 clear (not white) plastic drawer mat cover sheet from IKEA
j. 10# Slide binders.
k. 2# 450x370x25 foil faced insulation

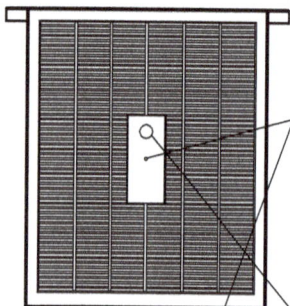

4 QUEEN EXCLUDERS (2 FOR PARTITIONS)

FOR ZEST HIVE ONLY
Partition boards with 6mm white plastic Cordek sheets push fitted into face of excluder and bolted through with a 25x6mm bolt and 2 penny washers.
The 20dia. hole in the partition to allow bee entry via a ramp onto pieces of honeycomb in an in-hive container holding feed.
Fill hole with wine cork when not in use

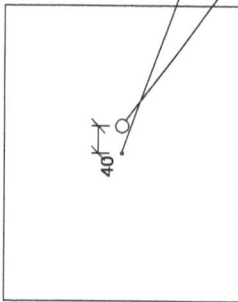

The ZEST hive customer may wish to dispense with the plastic carrier frame and buy 4 pieces of 69x20mm wood battens and support the frame end lugs on the insulation block walls. 22mm. holes will have to be drilled through the block walls to allow bee entry and ventilation. A virgin cage with a cap to be deployed in each hole to control bee access and ventilation

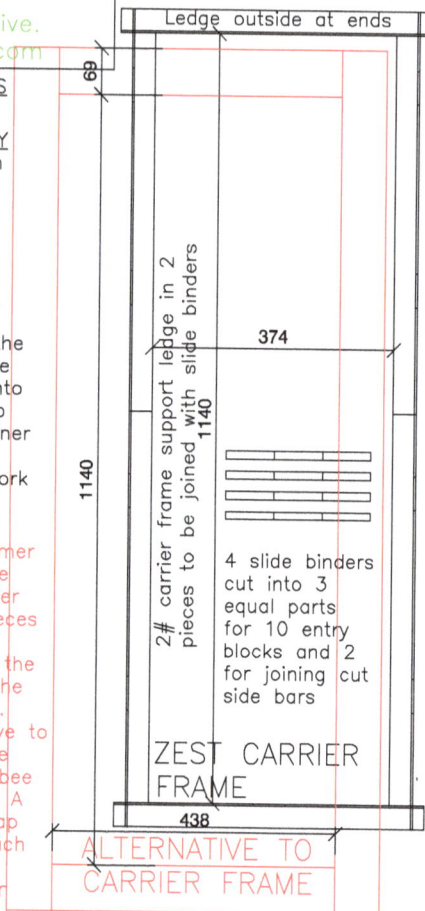

2 BOARDS
To convert 2 Queen excluders into partitions

Ledge outside at ends

2# carrier frame support ledge in 2 pieces to be joined with slide binders
1140

374

4 slide binders cut into 3 equal parts for 10 entry blocks and 2 for joining cut side bars

ZEST CARRIER FRAME
438

ALTERNATIVE TO CARRIER FRAME

17

Freedom Hive

Paula Carnell

I have kept bees for eight years in Somerset to learn about their life cycle and health. I have worked with bees across the UK and South Africa using a variety of hives including Langstroth, National, Flow, WBC, Warré, Golden, skeps, Sunhives, observation, freedom, log and Zeidler hives as well as unique custom-made hives for various clients. I currently regularly keep and manage a total of 25 colonies.

Freedom hives are based on a basic log hive design most commonly used in Europe.

The Freedom Hive has modifications which add insulation and a removable base and top from which the bees can be viewed or accessed. Weighing 17kgs it can be hoisted up into trees or placed on a tripod stand. The hives on tripods can have a straw hackle added to protect them from rain.

Matt Somerville creates his beautiful hives from the highest quality Western Red Cedar, being a light durable timber. Inside the walls is an insulated layer made up of wood shavings and compressed wood.

These hives when placed in trees are used to entice swarms that may remain for many years. Should honey need to be collected, a small skep 'super' or a box, can be added above the hive after removing the lid. A small entrance is exposed which allows bees to create comb and store honey above the main hive.

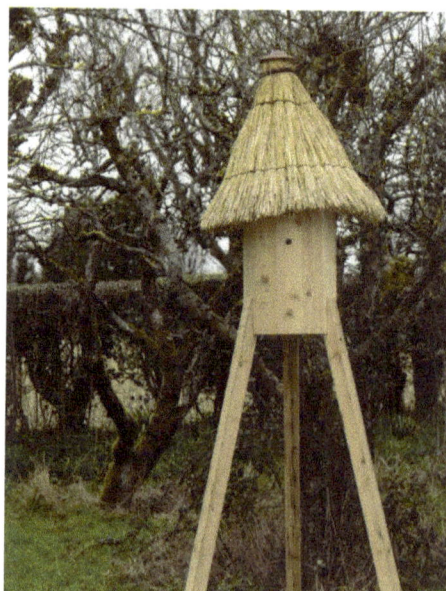

With three entrance holes in the main body of the hive, bees can enter and exit to access the active area of the comb inside. A new colony which is still building wax comb will primarily use the higher entrance. As the colony grows, bees use the lower entrances. The holes can easily be blocked by corks or adjustable plates although I found that the bees will block and open entrances with propolis as it suits them.

Freedom hives have now been placed in woodland and gardens all across the UK and Ireland and the majority attract swarms within a few days of them being placed in the Spring and Summer months.

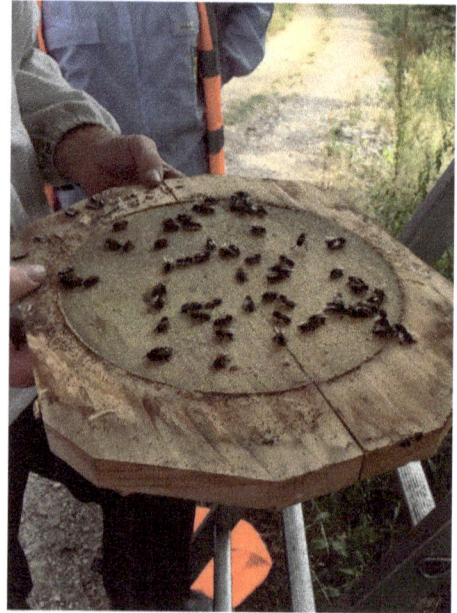

My first one was occupied by a swarm that was hanging off an extremely high and inaccessible branch for 10 days, during some dreadful wind and rainstorms in May 2017. The hive was positioned in a tree around 100metres away from the swarm and within 4 days the swarm had vanished, and the hive was full.

The colony can be viewed by removing the base, held in by screws, and looking up into the hive, the wax comb can be viewed and any debris on the base can be analysed.

Should a colony die, new colonies quickly move in and do not seem to be deterred by the work needed to clean out the hive. The benefit of comb already being in place is an obvious attraction and gives future colonies a head start should the summer be wet or cold.

The interior of the hive has been left with rough wood as Matt has found that the bees prefer to attach comb to a rough surface, this also makes it easier for the bees to protect themselves although of course it does mean that removal of individual combs for inspection or honey extraction is not recommended.

This is a hive for beekeepers wanting a permanent source of swarms or non-beekeepers wanting to provide a safe home for honey bees without interference and regular manipulation. Beginners and experienced beekeepers alike can learn a lot about the lift cycle and behaviour of bee colonies just from observing such a hive

The Freedom Hive on a stand could be used for pollination, placing the hive in orchards or within crops that do not have suitable trees to hang a Freedom Hive in.

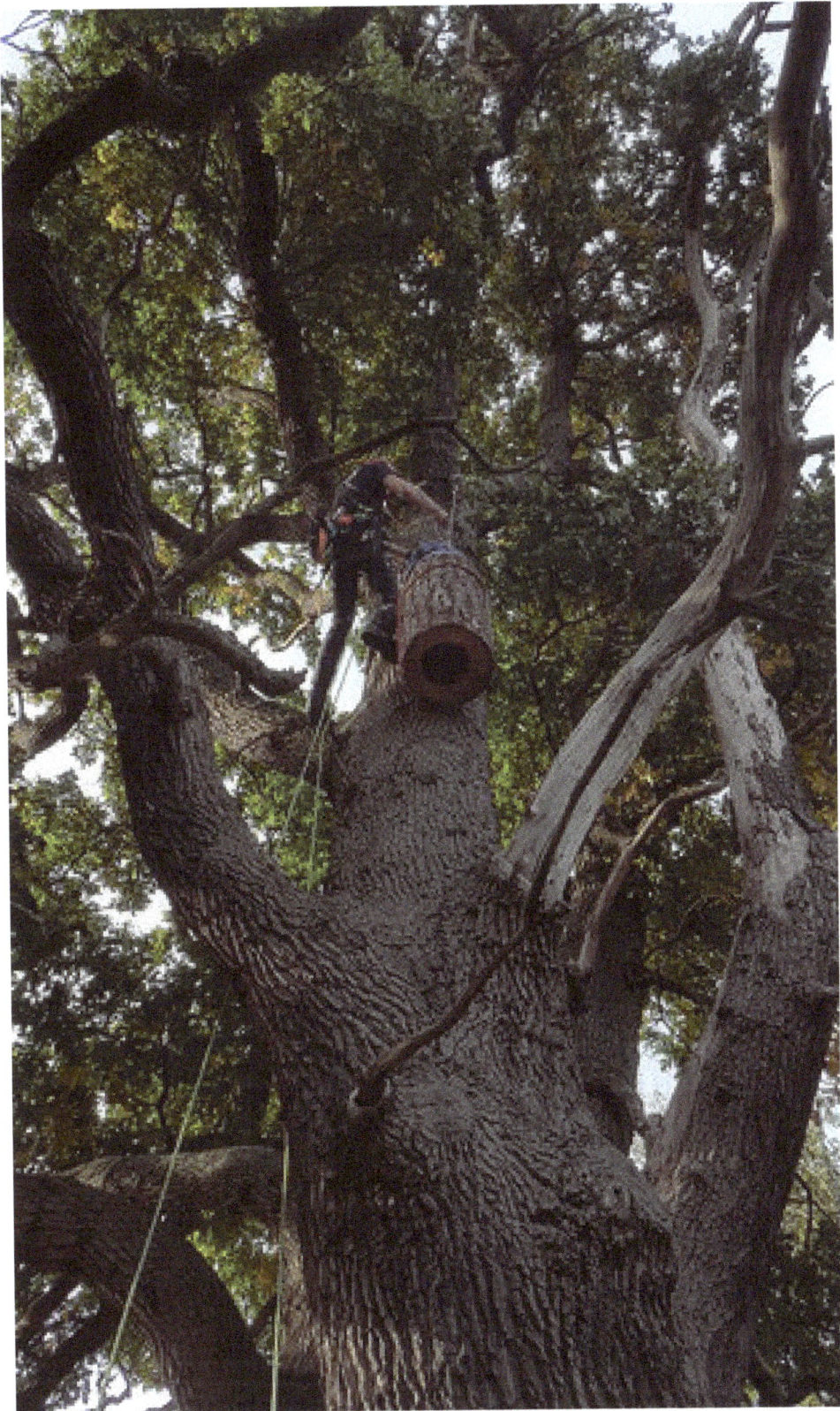

Advantages	Disadvantages
Suitable for anyone interested in keeping bees	Not for conventional honey harvesting
Minimum intervention	More expensive than a conventional hive, although reflected in the design and workmanship as each hive is handmade
Aesthetically pleasing	
Can be hung in a tree or freestanding	
Long-term low maintenance hive	

18

Sunhive

Paula Carnell

I have kept bees for eight years in Somerset to learn about their life cycle and health. I have worked with bees across the UK and South Africa using a variety of hives including Langstroth, National, Flow, WBC, Warré, Golden, skeps, Sunhives, observation, freedom, log and Zeidler hives as well as unique custom-made hives for various clients. I currently regularly keep and manage a total of 25 colonies.

I made my Sunhive in 2017 having first seen them on the Natural Beekeeping Trust's website and at Art in Action 2016.

It took us 3 days on a class to make our hives.

Using organic rye straw, grown in Somerset which was carefully stripped of seed heads and any stray grasses. We used provided formas to make our baskets the precise measurements required to fit the guidelines from Guenther Mancke's book 'Sunhive'.

The hive was based on an idea from watching the shape that bees create naturally when in a swarm or making a hive in a tree hollow or hanging from a branch. The first opportunity I had to house bees in my Sunhive was after collecting a particularly large swarm settled in long grass. I collected them into a skep them brought them back and positioned a ramp allowing the bees to walk into the hive.

This was quite straightforward and soon the bees were marching up the willow entrance into the hive. I scooped up some handfuls of bees and dropped them into the top entrance. The next morning, I found a large cluster around the base of the lower basket. This appears to be a common occurrence when introducing bees to a Sunhive.

Looking inside I could see that the old comb I'd placed in as bait had dropped from the frames and was blocking part of the entrance. Once removed the bees moved in.

Comb is drawn by the bees from the arched frames placed in the top half of the Sunhive. They rest on a flat wooden platform and can be removed easily by hand.

The idea is that the frames are sized so that any comb hanging from them can drop below the top basket and into the space covered by the larger lower basket. The lower basket can also be removed to view any inhabited colony or to inspect the comb.

Should honey be extracted it can be done by removing the outer frames which do not have any brood comb and then strained through cloths.

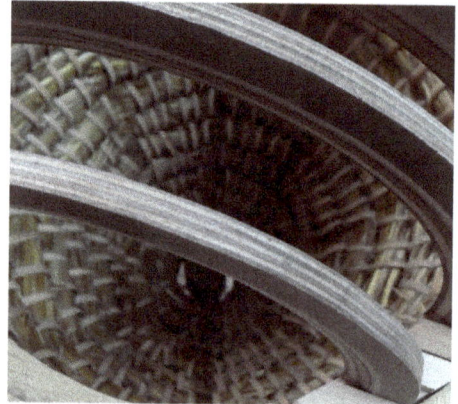

Advantages	Disadvantages
Pleasing to look at	No upper entrance/exit
Accessible frames that can be inspected and removed for honey or disease checks	Needs to be housed in a protected area or cloomed with cow dung or hackles to protect from rain and wind
Lower basket can be removed to observe a larger colony	Hand-made and so beyond many beekeepers budgets
Removal of the top wooden disk can enable access to the top of the frames to add food or a super to take honey from	Lower entrance may be too small and restricted to encourage larger colonies to move in
Suitable for natural beekeeping	Not easily transportable

19

Golden Hive

Paula Carnell

I have kept bees for eight years in Somerset to learn about their life cycle and health. I have worked with bees across the UK and South Africa using a variety of hives including Langstroth, National, Flow, WBC, Warré, Golden, skeps, Sunhives, observation, freedom, log and Zeidler hives as well as unique custom-made hives for various clients. I currently regularly keep and manage a total of 25 colonies.

Following on from work with Warré hives, Matt Somerville was also observing wild colonies living both wild in trees and in his Freedom hives. He designed the Golden Hive based on a Topbar hive, except with complete frames for the wax to be attached to. The basic brood box is deep at 56cm replicating the most common length of combs in the wild. As with horizontal topbars, the bees tend to keep their brood in the centre and honey stores are towards the outside.

Using a dummy frame, the hive can be restricted and expanded as necessary.

There is also a Golden Bait hive which I have had great success with. It can be placed high in a tree and with 8 of the golden frames, can sustain a passing swarm for several weeks or even months depending on weather conditions. In good seasons, bees have been able to create and fill one frame a week. The bait frame can be lowered, and the frames placed directly into a larger Golden hive in a permanent position.

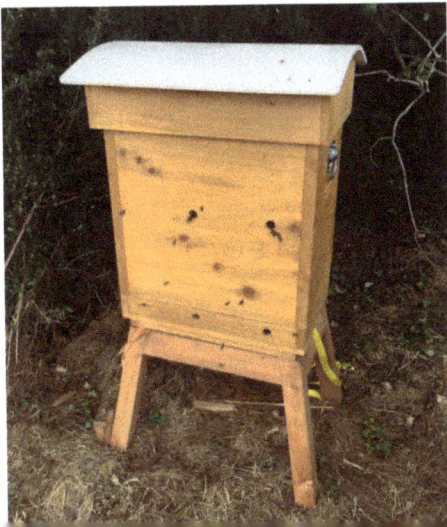

The hive is beautifully crafted in Western red Cedar with a base tray that sits on the stand. As in the wild, bees in trees would have a layer of old wax, and debris below their colony in which insects and microbes live and break down the combination of detritus. In my opinion this interaction and close proximity to the bees may aid the all-important microbiome which assists with natural immunity. A symbiotic relationship to the benefit of all concerned. The large brood box sits on top of the base tray, and is covered with a hessian cloth, an insulated eke, then a separate curved roof.

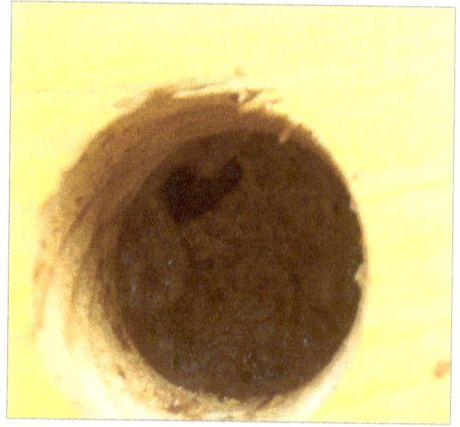

One ingenious touch is the frames having nails attached to create the beespace between them, both top and bottom.

There are three entrance holes in the brood box and two in the lower tray, they can be blocked by using an entrance gate or a cork!

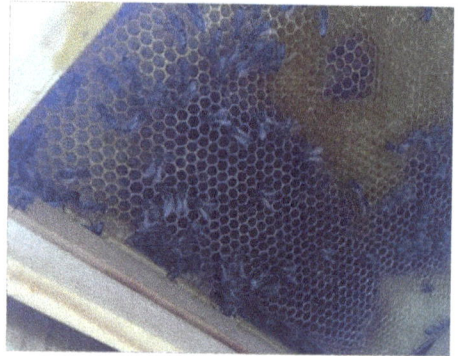

Pros

- Beautiful to look at and attractive to bees, swarms move right in!

- Easy to handle with full frames holding natural drawn comb

- A dummy board allows for extending or reducing the size of the hive

- Honey can be taken by removing frames with honey

- Handles on the side for ease of transportation

- Stand positions hive at a convenient height

20

Flow Hive

'Less labour, more love'

Lynne Ingram

I have been keeping bees in Somerset, for over 40 years. My apiaries are in cider orchards, and on an arable farm. I keep bees because they are fascinating creatures, and I never fail to be amazed at what I see when I open a hive. I currently have 25 hives, most of which are Nationals.

Imagine the scene – you've just finished extracting the last of your honey. You feel exhausted, and however careful you were, everything is covered in a thin film of honey.

- You have put on your suit;
- fired up the smoker;
- visited your bees;
- put on the clearer boards;
- gone back to get the cleared supers;
- carried them back to your car;
- carried them to your honey extraction place;
- cut off the cappings;
- spun the honey out in your extractor (thank goodness for that new electric extractor!);
- run the honey through a filter;
- emptied it into a settling tank,
- before bottling it.

And you still have to deal with the cappings; wash everything up; wash the floor and take the wet supers back to your hives for the bees to clear them.

Does this story seem familiar? Do you sometimes wonder if there is a better way of doing things?

Well you aren't the only one to dream of this. 10 years ago, Stuart and Cedar Anderson in Australia had extracted their honey and were thinking the same - but they decided to do something about it, and invented the Flow Hive.

The Flow Hive is probably the most innovative beekeeping invention since the discovery of bee space in 1851, and it promises a much easier way of harvesting honey – easier for the beekeeper and easier for the bees. It sounded very exciting when I first heard about it, so I decided to find out more.

I first saw details of the Flow Hive when an Indiegogo crowd funding campaign was started to raise money to develop a Flow Hive production plant. Initially Stuart and Cedar set out to raise £70,000, but it turned into the fastest growing crowd funding campaign of all time raising over £10 million – 147 times their initial target. The campaign included some wonderful video footage promising easy extraction directly from the hive – no taking off the supers, cutting off the cappings, or spinning out the honey. It sounded too good to be true – an absolute dream for the beekeeper – and I was hooked!!

Initially the Flow Hive was only developed for Langstroth hives, and I run National Hives, so I decided to wait until National Flow Frames were developed rather than having two systems on the go. I had to wait but in time I was the proud owner of a National Flow Super (however, I have since found that you can adapt the Langstroth flow frames to to fit a National hive).

So, let me describe the Flow Hive – what is it that makes it so different to everything else?

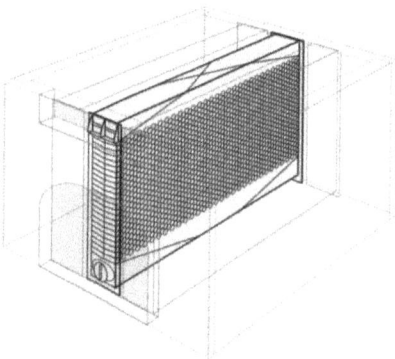

The Flow® frames fit into a standard Langstroth Super (8 or 10 frame). Two simple cutouts in one end of the super allow access for honey collection and end frame observation.

Flow frames

The key difference with the Flow hive is the frames. They are made of food grade BPA free plastic, and consist of partly formed honeycomb cells, with a split down the centre.

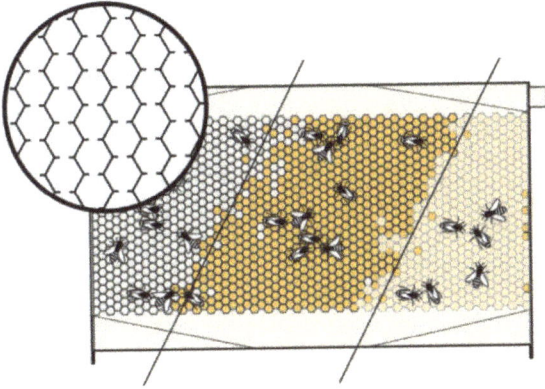

The Flow® frame consists of partly formed honeycomb cells. The bees complete the comb with their wax then fill the cells with honey, before finally capping the cells.

The bees complete the cells, fill them with honey and cap them. The totally ingenious design means that once capped, you can harvest the honey by inserting a honey tube in the base of a frame at the back of the hive and turning a key at the top of the frame.

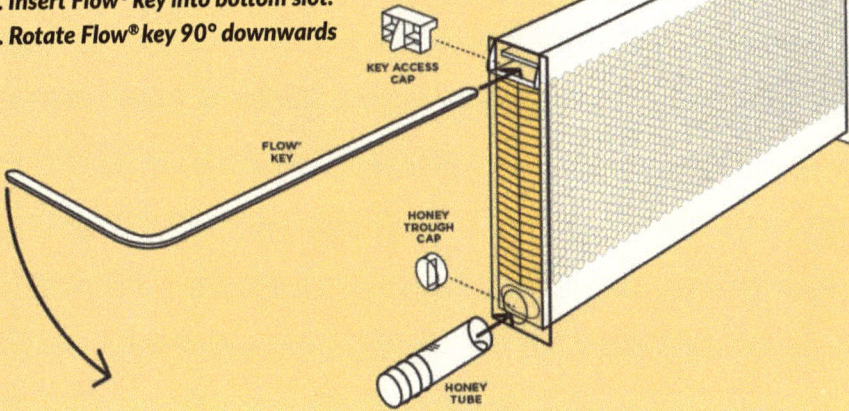

When the frame is full it's ready to harvest.
1. **Remove the key access cap and honey trough cap.**
2. **Insert honey trough into honey trough.**
3. **Insert Flow® key into bottom slot.**
4. **Rotate Flow® key 90° downwards**

KEY ACCESS CAP

FLOW® KEY

HONEY TROUGH CAP

HONEY TUBE

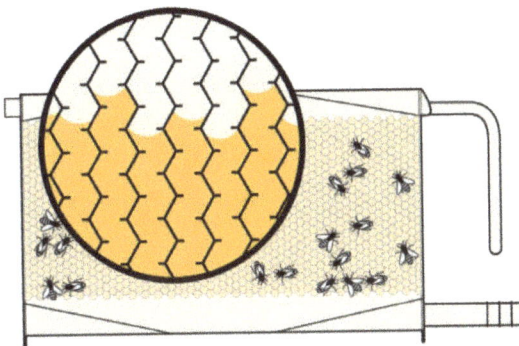

Inside the honeycomb the cells have now split and turned into channels for the honey to flow down. The bees remain undisturbed on the surface of the comb. If there does happen to be a bee down an empty cell it won't get injured as there is enough space between the comb walls.

This splits the cells under the cappings, forming channels down which the honey flows, into the bottom tubes, and then into your jars.

It's literally honey on tap from your beehive!

The bees remain undisturbed and continue to fly and forage while you harvest the honey. You can harvest one frame at a time – each one holding about 6lbs honey.

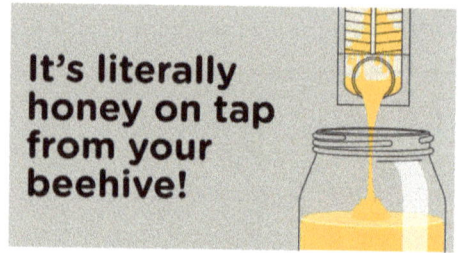

Harvesting one frame at a time means that you only need one Flow super, as you can harvest a frame when it is ready, and the bees can simply refill it. You can also keep the different flavours and colours of honey separate, as you can see what is inside through the ends of the frames. On the other hand, commercial beekeepers or those with many hives can attach pipework together so that honey from many frames flows into one big container.

Viewing windows

The second thing that differentiates the Flow Hive from others is the viewing windows. It's a bit like having a permanent observation hive.

There is one at the back of the hive, that allows the beekeeper to see the ends of the frames, and whether there is any honey in

them. This window also reveals the lower caps where the honey tubes for harvesting will be inserted. At the top of the back is a small removable piece where the flow key is inserted to open and close the frame channels in order to harvest the honey.

On the side of the Flow Super you will find another removable viewing window (with glass behind) that allows you to see the bees working on the end frame. This gives you an idea of how far the bees have got in the super.

So what was my experience with the Flow Hive?

After assembling the super (luckily the frames come ready assembled!) my first impression was how heavy the super was before even a drop of honey was in it, so I weighed it. A complete National Flow super with 8 empty frames weighs 11.5kg (25lbs 4oz). An empty National super and frames weighs 4.6kg (10.2lb). Imagine now that the bees have filled all the Flow frames with honey (8 frames x 6lbs) then you will be lifting up to 73lbs every time you do a colony inspection. Of course, the ideal scenario is that you extract your honey from the hive rather than lift it off, but the honey has to be capped, for that to happen. So, you either you wait until it is capped, lift up to 73lbs, or miss out on inspections.

I decided to use my super on one of my stronger colonies but had to wait until I was sure that the oil seed rape in the area had finished. I could not risk honey crystallising in the flow frames, as the flow feature would not then work.

Having got the super on, I regularly peered in through the windows rather than lifting out the frames and found that the bees were reluctant to enter the super. After some time however I started to see bees in the side window and knew that they had moved up. I could see that they were filling the gaps in the comb with wax and was looking forward to harvesting my honey the 'flow' way. I didn't however ever see any honey in the end of the frames, and eventually lifted out the frames to check what was going on. I found that they had stored honey in the centre of the frames, but it was not at that stage capped. Unfortunately, shortly after that inspection, the rain started and persisted for weeks. All the honey stored was eaten by the bees, so I didn't get to harvest any that year. But they had added wax to the comb, and they had stored some honey, so I was quietly confident for the following year.

Although in Australia beekeepers tend to keep their flow supers on all year, as they have warmer weather and a longer season, I needed to remove mine due to the ivy flow at the end of the season. Again, I didn't want to risk crystallised honey in the comb.

Year 2, and I decided to use the super in a different apiary where there was no risk of oil seed rape, so I could put it on earlier. However, despite a very warm summer, and a strong colony, I had little success. Again, the bees seemed reluctant to enter the flow super, even though there was wax on the comb. I had added a 'normal' super above, which they clearly preferred, and feel in retrospect that it would have been better to give them no other option but to enter the Flow super. Generally, I run my colonies on double brood boxes, but the Flow super is brood box size, so wonder if it would be better to confine the bees to a single brood box plus the Flow Super.

Having had 2 years' experience with the Flow Hive, I am looking forward to getting a good crop next year. I have learnt from these 2 years and plan to use the apiary with no oil seed rape, and to provide no other options for the bees.

Year 3, and I was determined to make this work, so decided to melt some beeswax and paint the combs with it. I then placed the Flow super on a colony which I kept on a single brood box, in an apiary with no Oil Seed Rape. Quite quickly they moved into the super and by the end of April I could see that some frames were almost full of honey. I was very excited. But a cool period of weather led to the bees consuming all of it before I could harvest any. A few weeks on and by mid July 3 frames were full and capped and I prepared to make my first harvest. **It was so easy!!!** I set up a small table behind the hive, and had jars ready to receive the honey. I removed the caps, put in the tube, and turned the Flow key. Within seconds clear warm honey was pouring into the jars. I needed to be quick to change jars as they filled, but averaged 6-7lbs from each frame. Each frame took only a few minutes to empty, and the bees took no notice whatsoever. **I am now even more of a convert!!!**

Within a week or so the bees had filled the frames, so I emptied them again, and the bees refilled them Altogether I had just over 100lbs of honey from the Flow super. I was also able to see different <u>coloured</u> honeys in the frames and was able to harvest them separately – more difficult to do in a 'normal' super.

Other peoples' experience

I talked to other Flow Hive owners in the UK. Some reported a good harvest from the Flow Hive with no special frame preparations. Others found that the bees were reluctant to enter the super unless the comb was painted with either melted wax, or a wax and honey mix. A short season is inevitable in Britain, especially if Oil Seed Rape is in the area. People reported using a 'normal' super while the Oil Seed Rape was out, adding the Flow Super later, and still getting a good crop. One commented that his bees entered the flow super 'if there was no other option'. For most people it took a while to get used to working bees in a Flow Hive. All however reported the ease with which they harvested the honey with no problems from robbing bees, although one person reported difficulties with wasps. It is essential to make sure that your receiving honey container is bee/wasp proof.

My verdict

The Flow Hive is one of the most fantastic innovations in beekeeping for over 100 years. It promises harvesting that is easier for the bees and easier for the beekeeper, and when all goes well it delivers on that promise.

It has been tremendously successful – there are now over 51,000 Flow Hives in 130 countries, and the second Indiegogo campaign for the improved Flow Hive 2 raised over £11 million GBP! It is however very expensive to buy. One Flow frame costs £66, with a complete Flow Hive costs £653 plus postage from Europe. (2018 prices)

There have been criticisms of the Flow Hive. It seemed so easy, that it attracted hundreds of people who had no experience of bees or beekeeping, but who wanted to 'Save the Bee'.

Initially there was no guidance for beekeepers, but they have now rectified this and there are beginners' beekeeping videos and information sheets on their website.

Beekeeping in Australia is different to the UK. They have lots of warm weather, and a long season. They don't have *Varroa* in Australia (as of 2018), and so the bees are not prone to the viruses spread by the *Varroa* mite. They do however have Small Hive Beetle, and American and European Foul Brood and so recommend 2 disease inspections a year. Apart from those inspections they only recommend inspecting every 4-6 weeks if you are in an area prone to bee disease. As mentioned before the weight of the Flow super means that it is difficult to manhandle when empty and even worse when full.

Overall, I would recommend the Flow Hive in situations where the beekeeper would struggle to find space for extraction equipment, and for somewhere to do the extraction, perhaps in urban situations. It would also be ideal for beekeepers that would like to be able to extract honey from a single source, or without the need for filtering or heating (truly raw honey). I would not recommend it for anyone who struggles with lifting – although Flow are almost ready to launch a new product to overcome the weight problem (Sept 2019).

My final word - I love the Flow Hive and am a total convert.

21

Darwinian Hive

Jeremy Barnes

I was born in Cornwall, my father's family is in Devon (originally based in Exmouth) and after a 42 year sojourn in Africa I now keep between 12 and 25 colonies in York County, Pennsylvania, primarily for my own fascination and pleasure; any honey (I have a sweet tooth) and enhanced pollination is a bonus. Besides six Darwinian hives I also have a top bar hive, some two and three queen castle hives, many bait hives and a suite of AJ hives in what was originally a chicken coop.

The term 'Darwinian Hive' may appear to be deceptive in that it might suggest leaving the bees alone and letting them 'evolve' to deal with current pests, pathogens and diseases. Far from it. The term was devised by Dr. Tom Seeley, until recently the Horace White Professor in Biology in the Department of Neurobiology and Behavior at Cornell University, and the author of several books on honeybee behavior, including *Honeybee Democracy* (2010), *The Wisdom of the Hive* (1995) and most recently, *The Lives of the Bees: The Untold Story of the Honey Bee in the Wild*. (2019.) He wrote a summary of his findings for the Nov 2018 issue of the *American Bee Journal*, and he is a keynote speaker at both the *Learning from the Bees Conference* in the Netherlands in late August, 2019, and at *Apimondia* in Montréal in early September, 2019.

Dr. Seeley's argument is that beekeeping looks different from an evolutionary perspective. Colonies of honey bees lived independently from humans for 80 million years or more, during which time they were shaped by natural selection to be skilled at surviving wherever they lived in Europe, western Asia, or Africa. Ever since humans started keeping bees in hives, some 5,000 years ago, we have been disrupting that fit between honey bee colonies and their environment by moving colonies to geographical locations to which they are not well adapted, and by managing honey bees in ways that provide us with honey, beeswax, propolis, pollen, royal jelly and pollination services but that interfere with their lives.

David Papke and I were stimulated by feelings of frustration over the relatively high levels of winter loss in our apiaries. In 2018, the average winter loss level throughout the county was 34%. Our challenge was to provide honey bees with a nest that is a better fit to their environment, enabling them to live with less stress and better health, and to do so by adapting our current equip-

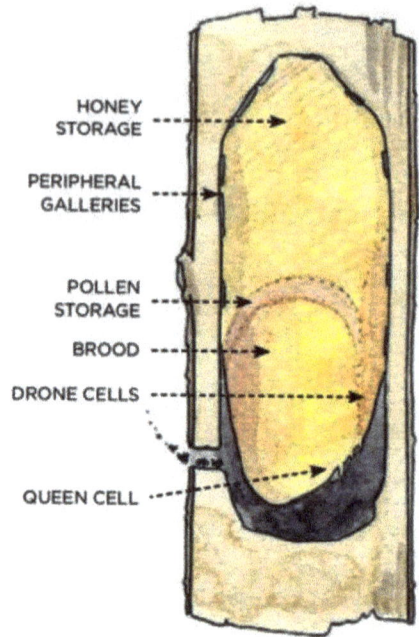

HONEY STORAGE

PERIPHERAL GALLERIES

POLLEN STORAGE

BROOD

DRONE CELLS

QUEEN CELL

The hive in its natural state. (Illustration after T.D. Seeley and R.A. Morse's The Nest of the Honey Bee, 1976.)

ment rather than buying anything new. The hives will be managed in terms of the practical suggestions described below, and we will measure our success rates in terms of winter survival rates, *Varroa* levels and the prevalence of pathogens and diseases,

When seeking a new home, bees look for a protected space, a small entrance and a well-insulated cavity. After they have discovered and moved into say a tree hollow, the worker bees will start drawing long, vertical comb from the top of the cavity for the queen to lay into, storing pollen and honey (which is converted into bee bread) around the central brood nest. When the bees feel the comb is getting too big to be stable, it will be cross braced in whatever way they see fit, with holes wherever needed to allow movement between combs. As the queen lays successive generations of brood, each one below the previous generation, the pupae in the cells above hatch and the vacated cells are packed with pollen to feed the brood below. When that pollen is converted to bee bread and is eaten, the cells are packed with honey and then capped. In this way, the pattern continues – a descending brood nest enclosed by a wreath of pollen and bee bread, topped by a growing dome of capped honey.

These feral colonies are normally small cavities (1.5 cubic feet) which are widely spaced from one another - at least half a kilometer apart. There is a small entrance (c.4 square inches) to each cavity, situated about 1/3 the way up the nest, the walls (ie. the tree trunk) are often 4 inches thick, the tree provides insulation below and above the nest, and the bees coat the sides of the nest with propolis.

The bees, which are genetically adapted to their location, are rarely disturbed, have a natural diet of diverse pollen sources and are not treated for diseases. The worker bees are the ones who select the larvae from which to raise queens, as much as 25% of the comb they build is for drones, and the pollen, honey, bees wax and propolis are not removed from the nest. The drones compete for mating and drone brood is not removed for mite control.

The bees are likely to abscond from the nest about every three years as the larval skins diminish the size of the cells, at which time wax moths clean out the comb, preparing the site for a swarm to move in and build new comb.

By comparison, we crowd colonies in apiaries, each colony in a box of 3 cubic feet or more, with thin walls of 3/4" on average, a large rectangular entrance of about 12 square inches, and the wood of the hive walls (normally white pine which has a low insulation value) is planed smooth, making it difficult for the bees to cover the surface with propolis.

The bees, which are not necessarily genetically adapted to their location, are frequently disturbed, fed artificial diets, and are exposed to homogeneous pollen sources as well as insecticides, herbicides and fungicides, not to mention chemicals for the treatment of diseases. Old comb remains in the hive unless the beekeeper replaces it on a regular two or three year cycle.

In terms of queens, beekeepers may choose the larvae for queen-rearing and even the drones for mating. Pollen may be trapped, honey harvested, and wax and propolis removed; drone brood may be removed and destroyed.

Our purpose was to design a Darwinian, or Natural, hive, using materials at hand and taking a bee-centric approach in the form of features that replicate a tree hollow and a management style that works *with* the bees for a mutually beneficial result.

Top box, inverted, before adding insulation.

We started with Langstroth hive bodies, partly because they are the most extensively used in the US, in particular the 10 frame mediums, with exterior dimensions 19 ⅞ x 16 ¼ x 6 ¾ inches and partly because we had many lying around unused that were ready to be recycled.

We insulated the two longer sides by gluing and screwing 1" thick timber on the insides, rough sawn so that the bees could cover the exposed surface with propolis. I chose so-called tulip poplar (actually a species of magnolia) because I had it available, it is light-weight, and survives well inside but not when it is exposed to the elements. The brood nest consists of two such boxes.

Top box, inverted, with insulation.

Insulating the two shorter sides is more difficult because the frames take up all of the inside space. It would be possible to add timber on the outside, but we chose not to do that. For winter purposes, I built a wooden shield to protect the windward side. 1" in depth, it has insulation between the wood and the hive bodies and is covered in tar paper to protect the insulation against the weather.

The interior dimensions of the insulated box are 18 ⅜ x 12 ⅞ x 6 ¾ inches, or .91 cubic feet.

Bottom box, with side and lower (not visible) insulation, and slatted rack. In this instance, the entrance is below the brood nest.

The medium box underneath the brood nest has insulated timber on all four sides, at least two inches of insulation material, a solid bottom board, a plywood cover (some insulation material is toxic to bees) and a home-made slatted rack using old cedar lath from demolished houses. Initially we included a circular entrance some 4 square inches in area as part of the bottom box, although I later moved that up to the first super, and am experimenting with a 3" shim that includes the entrance and which can be placed between the brood boxes, or between the upper brood box and the first super - feral hives seldom have the entrance at the bottom of the nest.

The assembled hive body, with a deep brood box and medium super, both of which are insulated on the longer sides.

The supers also are insulated as for the brood area, and the first super has the entrance. The upper cover, above the inner cover (or crown board) is similar to the bottom board - a medium box insulated on all four sides, with plywood and a space between the inner cover and plywood for ventilation purposes. There is also a screened ventilation hole in the back of the box.

The frames and foundation are as per normal usage, although we are experimenting with 16 inches deep frames that will cover both brood boxes without interruption between the foundation, much as the bees would build it in a feral space.

Curiously, Langstroth provided for insulation in his original design. He wrote in his patent application of 1853, referring to a double layer of glass as the hive walls: *"A very great degree of protection is given to the bees against extremes and sudden changes of temperature. The space of perfectly confined air between the double glass is some of the very best non-conductors of heat ... I have found the bees to be better protected in such hives than in any other unless where the wood is very thick, or doubled, and then the hives are clumsy and expensive... Small coves of wood may be placed against the posts leaving about half an inch space between them and the glass. This space, for greater protection, may be filled with tow, waste cotton, or any good non-conductor of heat".*

Tom Seeley describes five Natural Beekeeping Principles which are applicable to this hive design.

1. Wax comb is lipophilic and accumulates any toxins with which the bees come into contact, accumulating over time as the bees work that comb. By allowing the bees to draw their own comb, which he describes as their 'womb, home and larder.' rather than being forced to use plastic, pre-made or re-used comb, the bees ensure that the queen is always laying into fresh, virgin comb, which in turn impacts colony health. With naturally drawn comb, cell size will vary according to their needs (drone cells need to be bigger, for example) and the bees will draw comb at different speeds and sizes, according to the season's attributes.

2. Comb renewal happens naturally in a wild hive through the intervention of wax moths after a colony has absconded, whereas in conventional beekeeping the bees are stuck with the fixed comb they're given, re-using it as many times as the beekeeper sees fit – both for brood rearing and for honey storage.

In natural beekeeping, allowing bees access to fresh comb cleanses the hive, limits toxin build-up (both for the larvae and in the honey that is harvested) and means the hive is as healthy as possible, even in the face of widespread environmental chemical use.

Comb renewal can be encouraged by nadiring – placing an empty hive box under the existing colony, thus providing the bees space to grow downwards with minimal disruption.

3. Natural reproduction happens on two levels, either asexual (reproduction of the super-organism as a whole, when the colony splits via swarming) or sexual (when the existing queen is old, sick, injured or laying poorly, the bees raise a new queen who mates with local drones.)

Asexual reproduction allows the bees to choose both the timing of when a new queen is produced and her genetics. In the process of conventional re-queening, a commercially produced queen is typically mail-ordered and then added to the hive. Typically, this queen is bred for various traits – color, calmness, honey production - whereas in a natural system the traits may include disease resistance, strong comb-building abilities, strong foraging and successful swarming. The emphasis is on natural, locally adapted genetics as vital for the health not only of the individual colony but also for the 'super-super-organism,' meaning the community of bees in a given area, all interacting and exchanging locally adapted, resilient genetics to ensure long-term colony health for all.

4. Honey and pollen are both food and medicine to bees, which require a complete diet throughout their life.

The thousands of species of beneficial microflora found in both honey and bee bread are largely a result of the honey-making and pollen storage processes. As the bees collect, regurgitate and store nectar, layers of microflora are introduced. By the time the nectar becomes honey it is a probiotic powerhouse. Pollen is similarly processed inside the hive, becoming 'bee bread' with many species of microflora, which are passed via the feeding of larvae or the adult need for protein.

Conventional practices that harvest too much of the honey over the summer create a need to feed sugar water, which in effect is empty carbohydrates with none of the probiotics or medicinal aspects of honey the bees need, especially when they are stressed, cold or sick. In natural beekeeping it's essential to ensure the bees have enough honey stores at all times. In a crisis, feeding starving bees clean honey, preferably in the form of honeycomb and always from a disease-free hive, is the best substitute for their own honey stores.

5. Since a bee colony is a warmth organism, maintaining the colony's internal heat is crucial. Along with the nest scent of pheromones, maintaining this core temperature is intrinsic to colony health and disease suppression. A minimal intervention approach towards bee-keeping limits opening and disturbance of the hive. Minimising intervention doesn't mean ignoring one's bees; if anything, one's relationship with them is enhanced and more nuanced. Entrance activity, for example, can reveal much without opening the hive : bees bringing in pollen means the queen is laying well; bees flying in low and landing below the entrance might mean a good nectar flow with the foragers coming back full; frenetic activity in front of the hive might mean robbing in progress.

Just as the sounds and smells emanating from a hive can reveal much, so is nadiring (a term familiar to keepers of Warré hives, meaning adding supers below, rather than above, the brood nest) another way to minimize intervention.

Some practical suggestions to implement these principles and not included above are:

- **Work with bees that are adapted to one's location.** Local adaptation is a long-term investment and may not yield immediate results.

- **Space one's hives as wide apart as possible with entrances facing different directions.** Even placing colonies 30-50 meters apart reduces drifting and thus the spread of disease.

- **House bees in small hives.** Seeley recommends using just one deep hive body for a brood nest and one medium super over a queen excluder for honey. The honey harvest will be less, but disease and pest problems will be reduced. Certainly, swarming will increase, but this is natural and promotes colony health by suppressing *Varroa* mite populations.

- **Insulating the walls of the hives year-round**, not just in the winter.

- **Roughen the inner walls of hives or build them of rough-sawn lumber.** This will stimulate the bees to coat these surfaces with propolis, thereby creating antimicrobial envelopes around their nests.

- **Keep the entrance small** and place it either at the top of the brood area or at the bottom of the first super.

- Where possible, **positioning the hives high off the ground**, e.g. a porch or deck.

- **Allow the bees to draw out the comb**, using a starter strip on the top bar of each frame.

- **Recycle all comb frequently**, at the very most very three years, ensuring that the queen is laying into fresh, virgin comb, which makes a big difference to colony health.

- **Allow 10-20% of the comb to be drone comb**. This will improve the genetics in the area as only the strongest and healthiest colonies can afford to produce legions of drones. Drone brood also fosters rapid growth of *Varroa* mites, so careful monitoring of the *Varroa* levels is vital.

- **Minimize disturbances of the brood nest**. When working a colony, replace each frame in its original position and orientation. Also, avoid inserting empty frames in the brood nest to inhibit swarming.

- **Allowing bees to swarm and/or raise their own queens** means that the hive itself chooses the next queen. They choose the timing of when she is produced and her genetics v beekeeper re-queening (a commercially produced queen bee is typically mail-ordered and then added to the hive.) Instead of breeding for various traits that suit the beekeeper, the important traits favor the colony, as described in 3. above.

- **Use bait boxes** to increase or replace colonies and queens.

- **Ensure the bees have enough honey and pollen stores at all times**. In a crisis, feed starving bees clean honey, preferably in the form of honeycomb and always from a disease-free hive, preferably from your own colonies. Supplemental sugar provides empty carbohydrates with none of the probiotics, vitamins, minerals or medicinal aspects of honey that bees need, especially when they are stressed, cold or sick.

- **Minimize disturbances of nest.**

   (a) When working a colony, replace each frame in its original position and orientation and avoid disrupting the brood nest as much as possible.

   (b) Observe the entrance activity and learn to recognize different bee behaviors around the entrance, such as guarding, orientation flights and robbing behaviors.

      – Bees bringing in pollen means the queen is laying well and brood is present.

      – Bees flying in low and crashing onto the bottom board (or below it and crawling up) can mean that there is plenty of nectar around and the bees are coming back fuller than full.

      – Bees landing light on the bottom board can mean that nectar foraging is sparse.

   (c) Different smells and sounds are also indicative. Observation using all of the senses is key.

   (d) Nadiring.

- **Minimize relocations of hives**. Move colonies as rarely as possible. If hives must be relocated, move them when there is little forage available, or over winter when the bees are mostly inside.

David Papke, my colleague who was the main inspiration for the hive described above, and I built and populated ten Darwinian hives in the spring of 2018. At the end of the their first winter the survival rate was 83%, which is significantly higher than in the previous ten years with conventional hives, but it is a small sample over a short period and we intend to monitor closely their progress through the current year, including monthly tests of mites, pathogens and diseases.

22

Thermosolar Hive

Kathy Lovegrove

I took up beekeeping in 1999, after becoming aware of the steady decline in pollinators visiting my garden. My particular interests are in honeybee health; education and training; and queen-rearing for a more stable UK population of Apis mellifera.

I have responsibility for about 50 colonies in the Plymouth area using national hives (standard and 14x12), commercial and top-bar hives. In March 2019 I was invited to participate in a project to evaluate thermosolar hives in the UK.

Introduction

The thermosolar hive was patented in the Czech Republic in 2015 by Roman Linhart after ten years of research into ways of using heat to kill *Varroa* mites without harming honeybees. In its simplest form, the design is based on vertically stacked boxes for brood and honey; moveable frames; and a mesh floor for monitoring *Varroa*. However, virtually every element of this hive's construction has been adapted, using modern thermal technology to provide a more suitable year-round environment for the bees as well as facilitating the thermotherapy treatment.

Scientists have been aware of the prospect of eradicating *Varroa* mites with controlled heat treatment since the 1970s. The problem since then has been to develop a practical application of the concept. You are probably already thinking of the dif-

1: Thermosolar Hive - front view and panels

ficulties and risks associated with such a process. Several pioneers have applied themselves to these issues and developed different systems but beekeepers have yet to be convinced that the ideal solution has been found.

Roman Linhart's hive design was tested by researchers from Palacky University in 2015 for its effectiveness in destroying *Varroa* mites. They concluded that it was a highly promising method of eliminating mites without the need for chemical treatment, and that it was a significant advance on most practical applications to date.

The first thermosolar hive was brought to the UK by Tom Worsley in 2017 and installed at the Southill Energy Community's solar farm in Oxfordshire. I am involved in a project to test ten further hives at various sites in England. There is a lot more research to do on the practicalities of operating this system and the effects on the bees but here is a summary of the current state of play.

Thermosolar Hive design

The outer cover →

Thermosolar ceiling →

Thermosolar boxes →

Varroa mesh bottom →

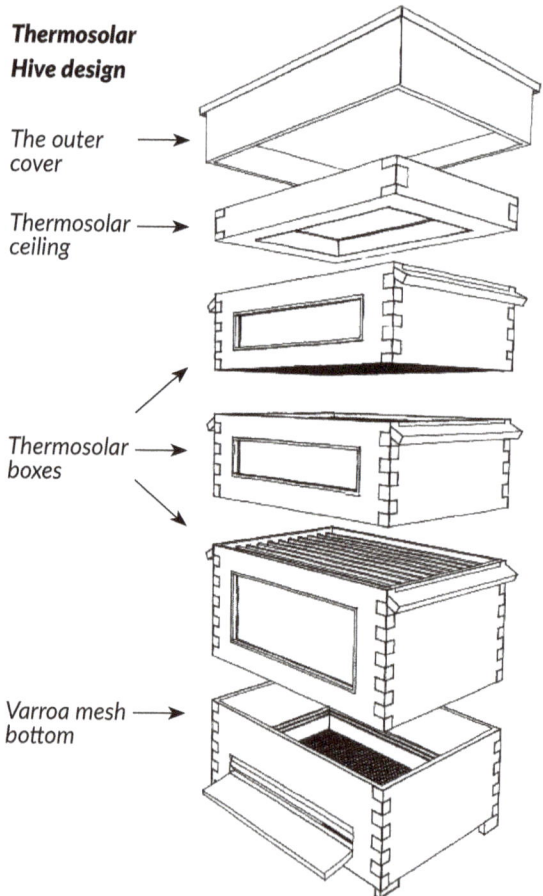

Hive Design

The roof (outer cover) is very well insulated to reduce the heat loss from the colony under A perforated metal sheet shields an insulating layer of fibrous material and allows the moist air to leave through four side vents (photo 2).

Another important function of the roof is in regulating the hive warming process during the periodic *Varroa* treatment. When this is carried out, the roof is removed and a solar panel (thermosolar ceiling) set in its place. Two thermometers are fixed through the front wall of the thermosolar ceiling, with the sensors inserted into the wax at the top and bottom of the central frame of the brood chamber (photo 3). Digital readouts are attached magnetically to the outside. Once the required temperature is reached the roof is returned to enable the internal hive temperature to reach an equilibrium.

The brood chambers, supers and frames used in The Czech Republic are Langstroths but the UK version is deep National 14 x 12 brood box with National supers. The latter are too shallow to take front thermosolar windows as shown in the diagram but this does not seem to be a significant disadvantage. At the moment the boxes only hold 8 frames plus a 'dummy board' but an 11 frame model is apparently in the pipeline. The frames run 'the warm way' so access to the bees is from the rear. All boxes have substantial handles and are fully insulated.

The key feature of the brood chamber is the thermosolar 'window' in the front. The advice is to site the hives facing south-east as this 'window' is glass-fronted but backed by a specially coated dark metal sheet to absorb the warmth of the morning sun. A white insulated 'dummy board' is also supplied which is placed between the internal wall and the brood frames to act as a heat diffuser and

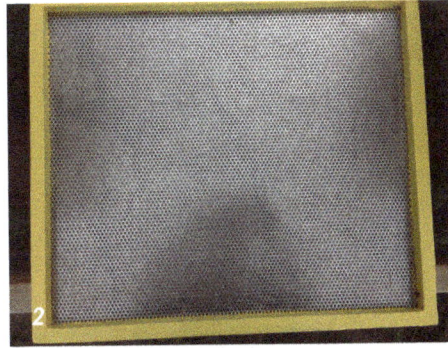

2: **Roof permo plate**

3: **Solar ceiling and super**

4: **Brood box and dummy board**

buffer against extremes of heat (photo 4). Inside the hive chambers, the walls are lined with black-coated aluminium which has both thermal and anti-bacterial properties.

The floor unit has three layers, each one with interesting design details. The upper section is made up of a slatted wooden 'false floor', a two inch (5cm) ventilation space and the main entrance. It is called a "building enclosure" on the website which is a bit of a misnomer since the wooden slats are there to discourage the bees from building wild comb below the frames. The slats are spaced so that the bees can still easily reach the entrance and they have a cooler refuge when the hive temperature becomes uncomfortable (photos 5, 6, 7).

5: Floor unit

6: Slatted false floor

7: Mezzanine floor

8: Rear view, floor open

9: Feeders unattached

In the middle of this bottom unit is a *Varroa* mesh floor which can be easily removed for cleaning, and at the bottom is an enclosure for placing a monitoring tray or pads. This section also has a bar with vent holes which are normally closed with wooden plugs. They must be opened during thermotherapy treatment, whenever the front entrance is closed and if the hive is being transported to another location. These vents can also be opened in early spring to aid ventilation if it is thought necessary to remove excess moisture (photo 8).

You will see a difference in the floor design between the exploded diagram and the photos since my hives have the useful alternative floor. This has a feeding extension at the back of the hive holding two 2.5 litre jars with percolating lids: for water and syrup (photo 9).

Aims of the Design

All the boxed elements of the hive, from floor to roof, are insulated to create a capsule with limited ventilation points, imitating the kind of site where bees would naturally choose to locate their nest. Using the principles of thermal inertia, the temperature change inside the hive is much less variable than outside and therefore more bee-friendly than other hives constructed with wood.

On a day to day basis it reduces the amount of time and energy which bees have to spend on thermoregulation, thus freeing them up to carry out more productive duties. Linhart claims that having a more stable temperature creates a more contented and successful colony overall. For example, he has evidence of increased honey crops, even double the quantity compared to conventional hives in the same apiary.

Linhart also maintains that spring build up is as much as two weeks in advance of other colonies because the reduced need to sustain a viable temperature in the cluster leads to fewer stores being consumed in winter. This means that foragers in early spring can concentrate on starting the new season rather than supplementing depleted stores to keep the adult bees alive and restricting the queen from laying too many eggs. The insulated environment boosts the warmth of spring days by extending the opportunity to sustain brood rearing temperatures in the region of 35C.

But the most radical function of the insulation is in eradicating *Varroa* mites from the sealed brood during the thermotherapy process.

Hyperthermia therapy

I have only had a chance to use the thermosolar hives for 6 mid-season weeks so I have not yet had to carry out this specialised course of action – although I seem to have done a version of it by accident! More of that in a later section. In the meanwhile I have looked into the theory behind it so bear with me while I share some of this with you.

You will probably be familiar with the use of relatively mild heat treatment for minor aches and pains but may not be aware that medical trials are being conducted where body tissues are exposed to excessive heat e.g. in order to damage or kill cancer cells, or to destroy Lyme disease microbes. Patients are prone to side effects but these are balanced against the positive results.

Using hyperthermia techniques on bees is based on a similar view of the good outweighing the bad. The therapy relies on the fact that bees are eurythermal organisms (i.e. they have adapted to tolerate a wide range of temperatures) whereas *Varroa* mites are stenothermal and can only survive within a limited range of temperatures. Thus the aim is to raise the temperature enough to kill *Varroa* mites without long term harm to the bees (photos 10, 11).

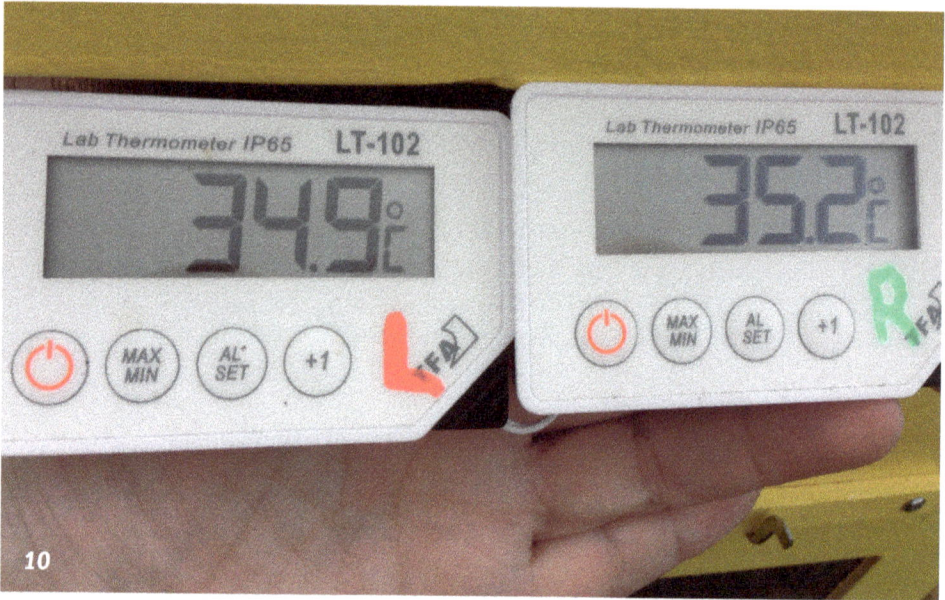

10

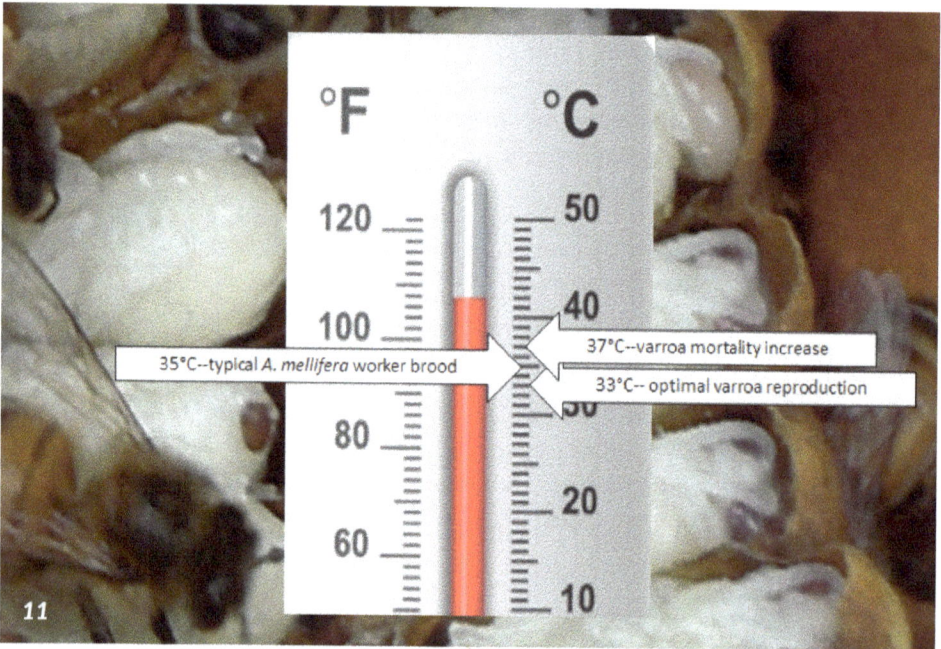

11

35°C--typical *A. mellifera* worker brood

37°C--varroa mortality increase

33°C-- optimal varroa reproduction

10: SF1 readouts

11: Ideal Varroa and Apis Mellifera.

The reproductive cycle of *Varroa* mites in the bees' active season lasts about 17 days with roughly 5 of these days spent attached to adult bees and the other 12 days in capped brood cells. Most chemical treatments are aimed at phoretic mites so have to be in the hive for at least one full cycle to have a significant effect and there is always that element of chance that some mites will not receive a fatal dose.

Hyperthermia therapy, on the other hand, is aimed at the mites in the sealed cells of pupating brood so only has to be applied on two specific days in the cycle. The first treatment eliminates the foundress mites and their developing offspring trapped in the sealed brood; the second treatment done 7-14 days later captures the remaining mites who have left the adult bees and entered the next batch of cells being capped. Theoretically 100% of mites in the hive during that period would be destroyed.

A word of warning: carrying out this treatment is not to be undertaken lightly. It is a far cry from, for example, inserting chemical strips into the brood chamber and removing them a few weeks later. The whole process takes 3-5 hours and has to be monitored throughout, just as if the bees are patients in a hospital. Beekeepers who use the thermosolar hive need to be willing to go to these lengths to protect their bees and/or to avoid the use of chemicals in the hive.

Carrying out the Treatment

There is plenty of detailed information available in the user's manual and on the website *www.thermosolarhive.com* covering all aspects of the treatment process but, for the purposes of this book, I will describe it briefly in the form of my plan for the treatment day which is coming up shortly in August.

1. Choose a warm, clear day with expected temperature of 20C+.

2. Reach the apiary about 9am.

3. Prepare the hives i.e. remove any supers; ensure suitable order and mix of brood frames; place sensors in central frame of sealed brood; space frames out at least 1cm apart; check clean *Varroa* tray is inserted; close all openings except a reduced main entrance.

4. Remove roofs to expose the solar ceilings.

5. Monitor the readouts. Observe the bees' behaviour or do simple tasks while waiting for the hives to heat up - or have a cup of tea and enjoy the scenery.

6. When the lower sensor reads 40C, start timing the 2 hour treatment period.

7. When the upper sensor reads 47C, replace the roof and allow the temperature to equalise throughout the brood box.

8. Observe the bees or otherwise amuse yourself while the 2 hours tick by.

9. When the time is up, ventilate the hive as much as possible.

10. When temperatures have returned to normal, reduce vents.

12: Theromsolar Hive – side / rear view and river.

What are the bees doing while this is going on? In short, experiencing heat stress and trying to cool the nest. Workers will disperse across the comb and engage in fanning to enable air to flow more freely. Older adult bees whose exoskeletal plates are harder than the younger bees cannot cope so well so some will encourage the queen to retreat to the mezzanine floor where they hang in festoons from the slats. Others will beard outside the hive or leave it temporarily. House bees will be regurgitating water or nectar droplets to cool the nest by evaporation and also be protecting the sensitive open brood by shielding them.

Bees are resilient and possess all these survival strategies whereas *Varroa* mites do not. Roman Linhart and other exponents of hyperthermia therapy have researched the appropriate levels of heat intensity and exposure time which will guarantee elimination of *Varroa* mites with minimum possible stress to the bees. Remember, too, that the technology used in the hive's design ensures gradual temperature change and diffused heat so bees can make these adjustments. It will be interesting to undertake the process myself and compare its effects on my colonies to that of other forms of treatment I have used.

Teething problems

When I first saw the thermosolar hive it didn't seem very daunting since it looked similar to conventional hives in structure. Also, the use of solar power meant that there was no additional electrical or mechanical equipment. Indeed, compared with other hyperthermia systems on the market, the thermosolar hive is relatively simple and compact. (Check out, for example, the Mighty Mite Killer, Bee Sauna, *Varroa* Controller and Vatorex).

13: Installing heavy hives.

14: Ooops!

Unfortunately, I was only given a verbal account of the hive's features and no manual was available until some time after I had transferred my colonies. (The translation from Czech to English was being improved.) So I didn't realise the crucial role of 'the dummy board' as a heat diffuser or that it had to be positioned behind the front solar window. I placed it behind the last inspected frame as normal.

Over the next few days, the bees were unusually bad tempered which was very awkward when they were meant to be the stars of a community event at the solar farm! At my next inspection I discovered that the sustained high temperatures of a heat wave and the efficient solar window had melted the nearest combs. I was shocked to think of the consequences the colony suffered due to my mistake (photo 14).

I was relieved when the manual did arrive. It is comprehensive, with details of the rationale behind each aspect of the hive's design and Linhart's first-hand advice on how to manage the equipment for optimum results. The website is worth a browse with interesting information and videos but I didn't find it very user-friendly.

The Future

There is a lot to learn about and from the thermosolar hive, and I feel privileged to have the chance to be involved in The Solar Bee Project. I would like to acknowledge the part played by Tom Worsley who introduced the hive to the UK, The Naturesave Trust who financed the equipment and Plymouth Energy Community who provided the site and set up the apiary. Once the bees have settled in at the various solar farms, it is hoped that the beekeepers will link up to carry out comparative research on how the design performs in each situation. There will also be many opportunities to open up debate on hive insulation and threats to pollinators amongst local beekeepers and the general public.

I think there is much merit in the insulating properties of the thermosolar hive and I look forward to testing these out. However, my main concern is for possible unrecorded effects on bees due to heat stress. Monitoring *Varroa* infestation is much simpler than detecting changes in behaviour, physiology or longevity in a colony of several thousands of bees. The research I viewed focused on *Varroa* extinction but also conducted checks on the bees over the next two weeks. The results reported that, if the hyperthermia treatment was carried out as instructed, there would be no adverse changes to the bees, wax or honey.

Plaque Naturesave

In conclusion, I am prepared to be convinced of the efficacy on *Varroa* mites of hyperthermia therapy in the thermosolar hive but much more scientific research on heat stress needs to be done to provide sufficient evidence that collateral damage is not an issue.

References

Randy Oliver's Scientific Beekeeping site. It was on this page *http://scientificbeekeeping.com/the-Varroa-problem-part-9* which was The *Varroa* Problem: Part 9 - Knowing Thine Enemy.

The exploded diagram came from the Thermosolar Hive website.

Bicik V., Vagera J., & Sadovska H (2016): *The effectiveness of Thermotherapy in the Elimination of Varroa Destructor* – Journal of Silesian Museum in Opava Acta Mus. Siles. Sci. Natur., 65: 263-269, 2016

23

SunNat and Dexter Hives

Roger Chartier

I have kept bees for 8 years and currently have 7 colonies which are kept in our garden and in an out apiary. My hives are Nationals as well as a variety of natural hives. The reason for keeping bees is to pollinate our plants

About 8 years ago, my wife Sue and I, both keen gardeners, decided we would like to keep bees to pollinate our plants. If they gave us some honey, then that was a bonus. So, we did a course, read endless books, purchased a National hive, a Nucleus with mated bees and off we went.

It was not very successful. Everything went wrong, pests, diseases, swarming. We were at the point of giving up when we went to a talk about natural bee keeping given by Heidi Herrmann, which changed our ways of supporting the bees. The essence of the talk was that bees know what to do, and it is our job just to support them. They should make their own comb, as and when they want to. Heidi had brought a Sunhive along to the talk to explain how it could be done.

The Sunhive was invented by Guenther Mancke to mimic the perfect bee nest shape which is the shape of a large avocado with a circular top and a pointed end. The hive has arched frames of various sizes to match the outside dimensions of the perfect comb, and the bees build their nest on these frames. I bought the Sunhive book, which contained all the dimensions and so decided to build my own.

You may well ask, why make changes to the original design? We came to the decision that the principles in which we wanted to keep bees, did not quite match the original Sunhive design.

Our Ethos is "*Bees develop in their own time and on their own comb*".

"*Beekeepers are able to inspect easily with the minimum of disturbance to the bees*".

Wooden Sunhive

The original Sunhive is made of straw covered in cow dung, with a protective roof. It has a domed roof and hangs 3 metres in the air. I believe we have a "duty of care" to our bees and the public, and because of this it is vital that the bees are inspected for pests, diseases and swarming. In the original Sunhive, it is difficult to inspect the individual frames. The curved frames are covered at the top with a cloth. This covering cloth has to be removed completely before any frames can be looked at. This means that all the hive is simultaneously exposed to the light and change of atmosphere.

I decided to make the new Sunhive out of wood and change the curved top of the frames to be straight and hang from top lugs like National frames do. I also realised that the horizontal dimensions of the Sunhive where similar to that of a National hive. Because of this, I was able to make the top piece convert from the circular Sunhive section to a square section the size of a National hive. The nest is the size and shape of the Sunhive but transfers into

the National size which will take standard super boxes, queen excluder and roof. I can also add a super above to collect any surplus honey not required by the nest. I can also feed in the normal way if necessary. The walls of the hive are 50mm thick and the funnel was turned on a lathe.

We collected a large swarm and put them in the top of the hive, and we packed the gap around the funnel with lettuce leaves and waited. Three days later the lettuce leaves wilted, and the bees found their way out. When we looked at the frames there was an incredible sight to be seen. The bees were hanging like a necklace linking legs and building their comb in mid-air before attaching it to the frame.

I have made a total of 4 hives, which we have been able to inspect the frames reasonably easily, with both of us being either side of the frame to view, as the combs are too large for easy manipulation. It enlightened us as to the way a true bee's nest is constructed as the bees build drone cells as they need, and queen cells hang like icicles on the sides. We have even been able to split the frames from one hive, which was just about to swarm, into another to start a new colony. I had to build a special tall nuc box, and it was very heavy!

I concluded that after using this type of hive for a few years it is brilliant if you start with a large swarm. There were issues which had to be resolved to make it more inclusive to varying swarm sizes. Because it has 5 separate frame sizes, it is not possible to use a "closure board" to make the nest area smaller. This is not suitable for smaller colonies starting out, as you can lose them because they cannot keep the large volume warm.

Guenther Mancke said bees do not like square corners and prefer curved edges. This may well be correct, but he did not take into consideration one key attribute of bees that they are ADAPTABLE.

If you have ever collected swarms, you will know the odd spaces that they will build their nests in. Since the ice age they have had to find somewhere to live in to protect them.

Dexter Hive

I wanted to be able to cope with the differing volume of colonies and be able to vary the width inside the hive. I selected one of the Wooden Sunhive arched frames and made duplicate copies of it. With the addition of 2 closure boards at each end, the nest could be expanded and contracted as required, and also split into 2.

I thought of Top Bar hives as a solution as they could do this, but I did not like its design being too wide and too shallow. The flat bars are inviting the bees to build cross comb. The Dexter Hive has vertical sides before the normal sloping top bar sides. In the top part sits curved frames (as in the Sunhive) for the bees to build their nest. I have found the standard top bar is too wide and is unwieldy to work with. The Dexter is not as wide, but deeper, and has a hinged pitched roof. The pitched roof enables a rapid feeder to be used easily, above the top bars which have a flexible feeder hole.

The hive has 4 entrances so the hive can be split, so two colonies can live in the same hive box. The frame width dimensions I have made the same as a National, so National frames can be use with a minor adjustment of adding a thin frame cover to stop bees escaping through the gaps. We have been successful using this by being able to introduce BIAS brood frames into a queenless hive. As they expand, they migrate on to the curved natural frames, and in time the imported pre-waxed frames can be replaced. Surplus honey is placed by the bees on the end frames and these can be stolen without disturbing the main nest. The amount of surplus honey production is not high because I think the horizontal distance from the centre of the nest to edge frames is too long.

There is a bee keeping adage that says, bees like to "go up" and in this hive they have to go sideways.

SunNat

The SunNat is a hive in which the bees build their own foundation but also "can go up".

The top box is a standard National brood box and is made up with standard frames, but these frames have curved inserts (as in the Sunhive) rather than pre-waxed foundation. This also means standard national frames can be included, for example from a Nuc box for migration to the natural frames. Below the brood box is a box of similar size with 2 of the 4 sides being sloped, this is to allow the bees foundation to be finished as the base of a catenary curve (like the bottom of an avocado) as in the Sunhive and Dexter hive. Above the brood box standard national equipment can be used. One advantage of this hive is that a beekeeper can try the method of natural foundation frames without the considerable effort of specialist construction of the Sun or the Dexter hive. The closure frames are 18mm thick, which not only block expansion, but also add extra insulation. Please note the pictures show the hive is bark lined, this is not essential, I am trying to introduce Pseudoscorpion *Chelifer cancroides* as a means of natural *Varroa* treatment.

Conclusion

Sunhive is grand and majestic, looks imposing and needs large colonies.

Dexter is self-contained, easy to manage and flexible. Ideal for someone who wants to keep bees in a natural way and does not want all the paraphernalia of standard bee keeping.

SunNat is for standard beekeepers who want to look after their bees in more natural way, still using most of their existing equipment.

24

Engineering in the Honey Bee's Relationship with the Hive

Derek Mitchell

This started out when faced with some contradictions to the physics I had learnt at school on a beekeeping taster day in 2010 as regards hive design. A quest to find out what the honey bees real requirements were and making a low heat loss hive lead to being a post graduate researcher at the Institute of Thermofluids, School of Mechanical Engineering, University of Leeds. The beekeeping has moved from collecting honey from hives to collecting data from artificial trees with currently about 10 colonies.

Engineering the future hive

To move forward we need to go back to what the honey bee really does with a hive, is it just a shelter or is much more? To answer that question, we need to think about what roles the hive fulfils when honey bees takes it over and what is that collection of insects trying to do. First some deeper thought and deeper terminology on the problem before applying increased sophistication to the solution.

The honey bee – a super organism with a subtle phenotype extended

The phenotype, the physical reflection of the gene, cannot be limited to purely the biological aspects of the organism itself, as the organism must impact and change the environment, as described in the laws of thermodynamics and mass and energy conservation. For many organisms the realm of their influence extends only a very small distance from the biological tissue of the animal and for only a small amount of time after its individual existence ceases.

However for organisms in which this impact and interchange goes beyond the usual, and the reach of the gene into environment is significant, then it is referred to as an "extended phenotype [1]. The classic cases cited are those of beavers flooding areas with their dams and termites creating tall mounds on the surface to change conditions of their underground nests. Perhaps because of the relationship with man, the extended nature of the honey bee has been overlooked, along with their natural nests that are usually hidden within a tree, more obscure and impenetrable than the dwellings of either beaver or termite. As a consequence honey bees nests have been viewed anthropocentrically as static containers of honey and brood. This is in contrast to honey bees investment in finding and selecting suitable nests. On occupation they invest much wealth and effort in excavating then coating the inside with propolis and constructing the wax comb internals. It could be argued that this extended phenotype is multi-generational as the investments are bequeathed by the departing swarming queen to her daughters. The high value of the bequest is evident because any abandoned nest is actively sought by new occupants.

We recognise the control of water by the beaver, yet dismiss the control of its extended phenotype by the honey bee and limit our perception with glib judgement, "does not attempt to heat the inside of the hive"[2]. Where the fluid the beaver controls is all too visible to our eyes the fluids the honey bee controls are more subtle and require a greater perception to be understood.

The Thermofluid adept

Thermofluids is defined as the study of heat transfer including phase change, fluid mechanics, and the combustion of fuels. The fluids here include gases such as air and liquids such as water, solutions and fuels.

Honey bee metabolise a fuel - the sugars dissolved in nectar and honey, change the phase of the nectar water content from liquid to vapour, engage in heating a nest that has non trivial thermal properties, actively exhaust, humidify, dessicate and recirculate the air within their nest. They balance the conflicting requirements of a constant temperature nursery at high humidity with the variable heat load and low humidity of desiccation of nectar into honey, for practical purposes a sugar refinery. Their survival and success relies on being adept at thermofluids, adjusting the input of energy to the demands of temperature control and water removal, while keeping an air transport system fuelled during the summer, building up fuel stocks for the winter shutdown, and keeping the premises secure. Having the behaviours, or what engineers would term control systems, to achieve this, makes the description "thermofluid adept" somewhat of an understatement. The honey bees nest is more sophisticated than the beavers dam as it does not just constrain the mass flow of a single fluid, but involves multiple fluids and their characteristics. A honey bee nest impacts the mass flow of air, water vapour, water liquid and carbon dioxide, while modulating their temperature, velocities and mixture ratios. In addition, we have the heat flux through the nest walls, in and out of the entrance as well as condensation and evaporation of water. In short, a honeybee and its nest, the extended phenotype, can be arguably said to be "all about the thermofluids"

The correct set of tools

With thermofluids being such a crucial part of the honey bee and its nest, it becomes clear that to understand the honeybees relationship with its nest, and by our actions to at least not hinder the activity, we should apply the relevant intellectual tools we possess. That we normally use these to analyse sugar factories, buildings or nuclear power plants, should not be seen as a barrier to our investigation. An oxygen atom about to oxidise does not ask each carbon atom if it is organic before determining how much energy to release. At the level of thermofluids, origins do not change the maths. The complexity and subtly of honeybees could involve diving into the extreme complexity and multiple degrees of freedom they control, and that remains a task yet to be embarked upon, but we should acquaint ourselves with at least the basic measurement at the grossest level of the flows of the fluids involved.

Three quantities define the rate heat, water vapour and air can flow in and out of the nest, per unit of the relevant impetus

- **Lumped thermal conductance** – this is the rate of heat transferred per degree of temperature difference.

- **Water vapour permeance** – the rate of mass flow of water per unit of water vapour pressure differential.

- **Air conductance** – the rate of mass flow of air per unit of air pressure difference.

The importance of these quantities is evidenced in the most obvious of honey bee behaviours. Lumped thermal conductance is the quantity being changed when honeybees cluster, so reducing the rate of energy loss. The spreading of propolis reduces the water vapour permeance, so that honey bees are not overwhelmed by the water vapour originating from the tree sap, while desiccating nectar into honey. Finally, where excessive air conductance exists honeybees have been observed to close holes and entrances with propolis.

The interplay of thermofluids in the honey bee nest and common misconceptions of them are laid bare when one looks closely at the age old beekeeper discussions of condensation, ventilation, and heat loss. The subtle interaction is often synthesised into saws and sound bites, which show the problems of a mixture of anthropocentric thinking, the ambiguity of the English language, lack of common scientific knowledge and that some areas of recent research need a wider audience. We will look at this common concern of beekeepers to show how applying modern engineering research produces a clearer path for progress

Modern engineering on an old controversy – ventilation and insulation

The drive to produce cheaper hives and yet higher honey production in the 1930s received a boost in the form of the 2nd world war, when ships being sunk, and demand for wood and wood workers for other purposes, made simple easy to produce hives into topics for lunchtime radio programmes on the BBC. While over in the United States, E. J. Anderson was conducting experiments to see if the common solution to getting rid of the condensation in thin walled hives by adding a top vent, made any other difference. He concluded "As far as heat is concerned it appears that little heat is lost through the use of a small top entrance in addition to a bottom entrance". What Anderson had done was to simulate a colony in hive by using a 15W bulb in the bottom of the hive and then observing the temperatures both with, and without, a vent in the top of the hive in addition to the open entrance at the bottom, using just a single thermometer. Anderson drew his conclusion from the thermometer reading being unchanged. From his simple experiment has sprung the justification to top ventilate honey bee hives in all weathers and in all configurations. This work became important as it often quoted in the following years [3], and top ventilation became part of beekeeping and bee research orthodoxy [4].

Figure 1

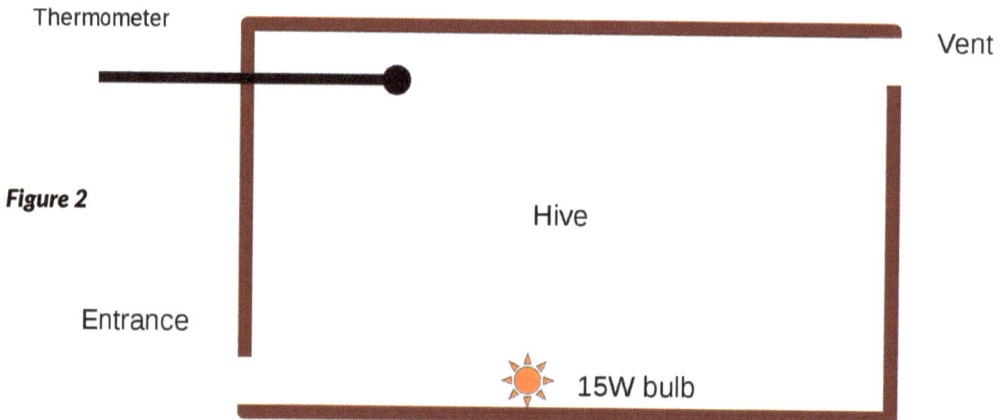

Figure 2

By the 1990s building science, had progressed to producing simple mathematical models of rooms with a heat source and a variety of ventilation schemes. The classic work on this was produced in 1990 by P.F. Linden [5]. This would explain that the light bulb was producing a jet of hot air that rose and created a pool of warm air at the top of the room. Any vent at the top would then empty that pool. The depth of the pool would then be dependent on the relative sizes of the top and bottom entrances. Thus, the simple experiment of Anderson seemed to be a smaller scale, simpler version of the detailed verification work carried out nearly 50 years later. However, the work in 1990 assumed zero heat losses (adiabatic) and that of 1943 had omitted the effects of changing the amount of heat loss, which would occur primarily through the roof.

In 2012, Linden's work was extended by Lane-Serff [6] to take account of the effect of heat loss and what might happen if the heat loss was changed. Lane-Serff explained that the magnitude of the flow out of the vent was dependent on the energy left after heat losses were accounted for.

Now we could explain what Anderson saw in 1943 in his limited experiment and then draw more accurate and wider ranging conclusions.

Apiarist Anderson's experiment through research engineer Lane-Serff's eyes.

If we use Lane-Serff model we can see that a hot jet is formed and so is a heat pool of almost uniform temperature (figure 3), with the heat leaking out through the high thermal conductance wooden roof. When the vent is added (figure 4), some of the warm air leaks out of the vent and the rate leaking out the roof is reduced slightly, leading to the warm heat pool becomes a bit shallower. The total rate of heat being lost does not change much. The temperature at roof level doesn't change much either. This is explained by the fact that the roof conductance is so high that the heat pool is not at a high enough temperature compared to the outside to drive a strong flow out of the vent.

Figure 3

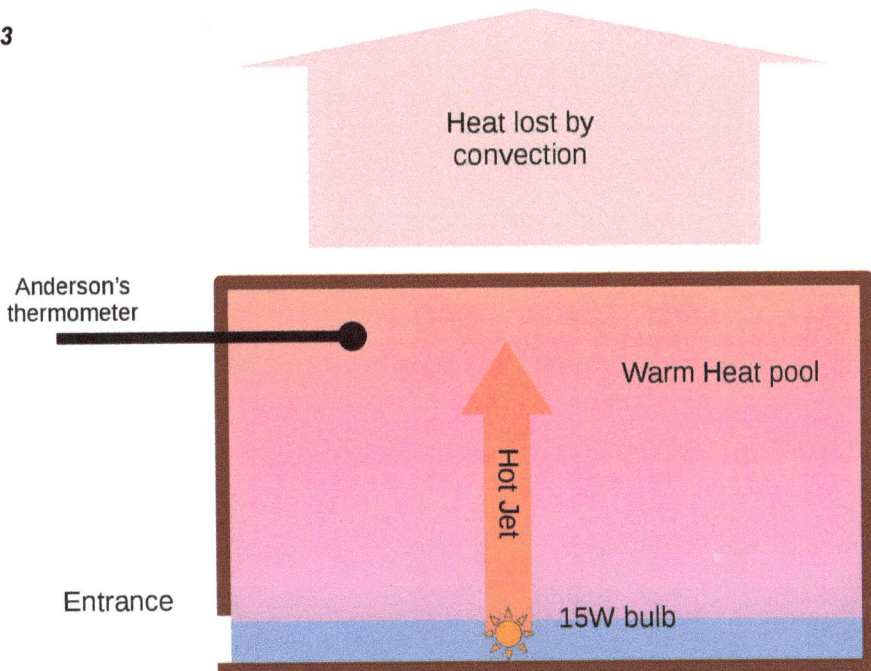

Heat lost by convection

Anderson's thermometer

Warm Heat pool

Hot Jet

Entrance

15W bulb

Figure 4

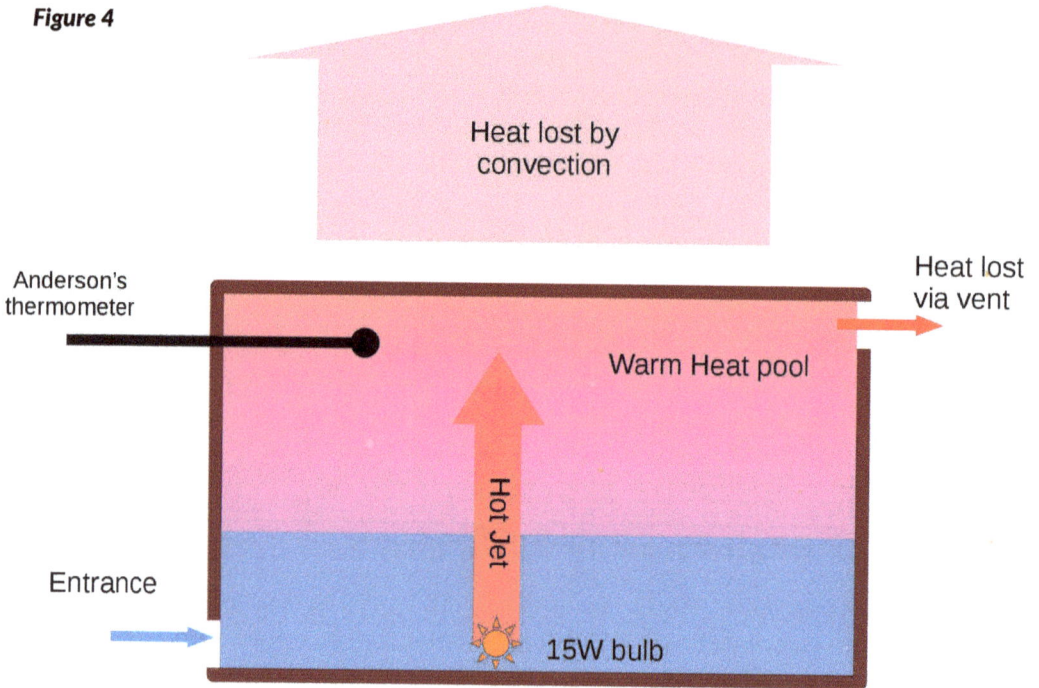

Heat lost by convection

Heat lost via vent

Anderson's thermometer

Warm Heat pool

Hot Jet

Entrance

15W bulb

Now consider the low heat loss case. With no vent (figure 5) The heat pool is only losing a little heat per degree K out through the roof, the temperature inside becomes higher.

Figure 5

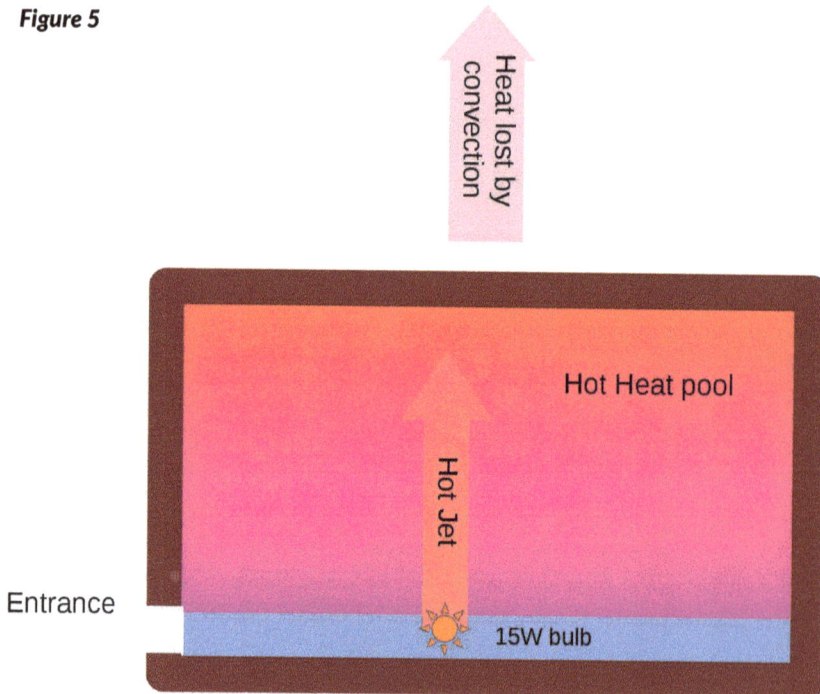

Heat lost by convection

Hot Heat pool

Hot Jet

Entrance

15W bulb

When the vent is added (figure 6) there is a greater temperature difference which gives a greater flow and hence a greater heat loss from the increased flow and temperature. The increase in heat loss is now substantial and the strong flow of hot air would now start to not only make the heat pool much shallower but also evaporate a lot more water from the hive internals and stress honey bees housed in such a situation, as found by Dodologlu in 2004 [7].

Figure 6

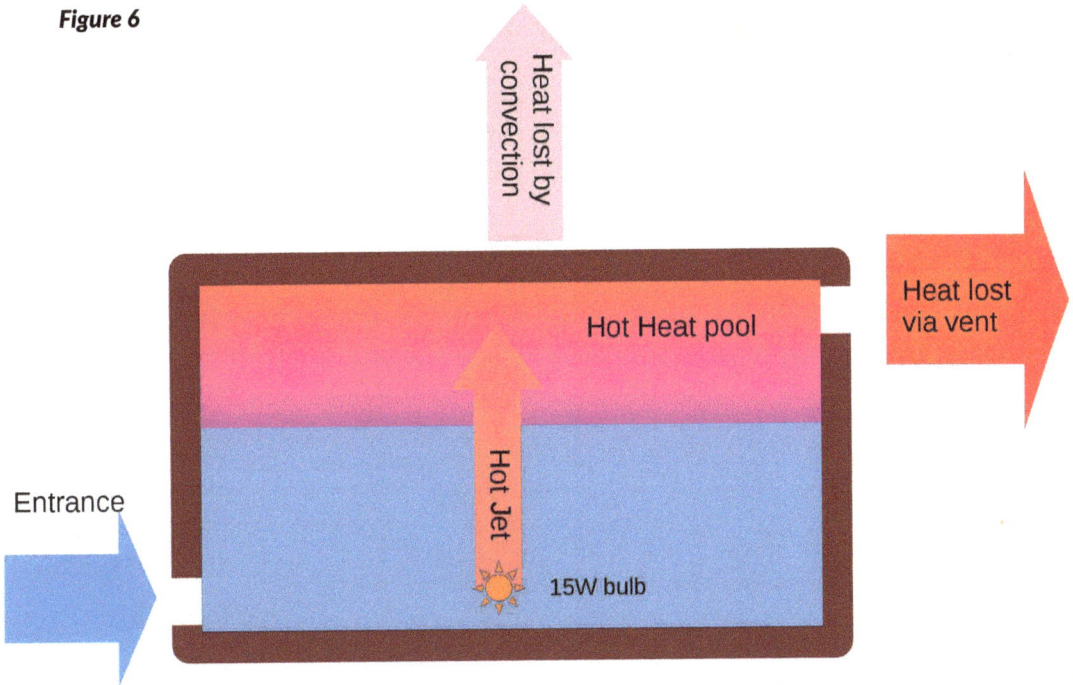

The Canada paradox and the Brother Adam effect.

How does this account for beekeepers in cold Canadian winter with a top vent in their well wrapped and insulated winter hives having success [3], while Brother Adam when he tried out insulating Canadian style in Devon UK having failures? [8].

The key here is to think what determines the amount of heat loss. It can be calculated as a product of the conductance and the difference in temperature difference, or the ratio of the temperature difference and the thermal resistance. This means that a well wrapped Canadian hive in a very cold Canadian winter is still having high heat loss and will have low vent flows so will not be badly affected by the top vent. However, a Canadian style hive in a mild Devon winter will have low heat loss, and therefore has high vent flows and the poor consequences.

Honey bees, in their natural home, a tree hollow, possesses insulation far in excess of Canadian beekeepers beehives and so given the work of Lane-Serf, honey bees would be in peril if they chose top entrances. Perhaps it is then no surprise than honey bees, when given a choice chose a bottom entrance nest [9]. They have through natural selection encoded the optimum engineering solution into their genes.

Conclusion

Genetics is information encoded in chemistry to provide a blueprint for an organism i.e. a genotype, the phenotype is its expression in the physical world. Biology is an organisation and structure of physical phenomena related to living things and is not something separate from physics. By understanding the physical processes in the extended phenotype space, we can understand the reasons why a genotype that encoded them prospers. Thus by using our knowledge of engineering and physics we can come to a greater understanding of why the honey bee is so successful and ensure that we can develop future hives that add to that success rather than ones that the honey bees tolerate and that are solely for our convenience. The honey bees have had millions of years to select optimal solutions for their extended phenotype and we are at only at the beginning of understanding exactly what those solutions are.

References

1. Dawkins R. 1982 *The Extended Phenotype*. W. H. Freeman & Co, Oxford.

2. Farrar CL. 1947 *The Overwintering of Productive Colonies*. In *The Hive And the HoneyBee*, pp. 425–451.

3. Currie RW, Spivak M, Reuter GS. 2015 *Wintering Management of Honey Bee Colonies*. In *The Hive And the HoneyBee*, pp. 629–670.

4. Ratnieks F. 2016 *Autumn Preparation of Hives for Winter*. See *http://www.lasiqueenbees.com/how-to/autumn-preparation-of-hives-for-winter* (accessed on 23 February 2019).

5. Linden PF, Lane-Serff GF, Smeed DA. 1990 *Emptying filling boxes: the fluid mechanics of natural ventilation*. J. Fluid Mech. 212, 309–335. (doi:10.1017/S0022112090001987)

6. Lane-Serff GF et al. 2012 *Emptying non-adiabatic filling boxes: the effects of heat transfers on the fluid dynamics of natural ventilation*. J. Fluid Mech. 701, 386–406. (doi:10.1017/jfm.2012.164)

7. Dodologlu A, DÜLGER C, Genc F. 2004 *Colony condition and bee behaviour in honey bees (Apis mellifera) housed in wooden or polystyrene hives and fed 'bee cake' or syrup*. J. Apic. Res. 43, 3–8.

8. Adam B. 1975 *Beekeeping At Buckfast Abbey Autumn & Winter*. In *Beekeeping At Buckfast Abbey*, pp. 55–58. Northern Bee Books.

9. Seeley TD. 2010 *Honeybee Democracy*. Princeton University Press.

25

Developments in Hive Components

1: Insulation and hive materials

Insulation

Basics

The fundamental principle of Thermal Insulation is that less dense, lighter materials contain heat best. Also the thicker the insulation, the more effective will it be. If gasses such as air are being used as insulators they will work better if convection currents are eliminated by trapping the gas in closed cavities, cells or foams.

Environment

The temperature differential between the outside and interior and the amount of wind and rain impacting on the hive are key factors. For materials to work at their best they must be kept dry. Wetting not only increases the conductivity of saturated materials, its evaporation causes cooling and increases the temperature gradient - this is how a fridge works. So, locations exposed to the wind will require more effective insulation, protection from wind-blown wetting and consequential cooling.

Materials

The table below[14], derived from various advisory sources, shows how commonly used materials perform, from tough ones that conduct heat to soft insulations that do not.

Predictably the densest materials transmit most heat and the those that are least dense and trap air to stop convection currents perform best.

The rate of heat loss is expressed in Watts per metre of thickness of the material for each degree Kelvin of temperature difference across the material. (Centigrade can be substituted for Kelvin in these figures.) The rate of heat loss decreases in direct proportion with the thickness of the material and increases proportionately with the increase in temperature gradient.

The total heat loss across a panel comprising layers of different materials can be calculated by taking the resistivity of each material (which is the reciprocal of the u value constant in the table below) and multiply this by its thickness. Do this for each layer, add them up and this is the R value (resistance) for the panel. The reciprocal of this will then give the U value for the particular panel. If this is multiplied by the panel area and by the temperature difference between inside and outside, this will give the heat loss through the panel in Watts. If this is done for hive walls, roof and floor, without ventilation, the total heat lost - and therefore the heat the bees must generate to maintain the internal temperature - will be the sum of these figures plus allowance for convection and thermal bridges such as battens or metal fixings which tend to increase heat loss.

Material	Density	Heat loss(U value constant)
Metal sheet (steel)	7800Kg/m3	50.000Watts/metre K
Concrete and stone slabs	2400Kg/m3	1.930Watts/metre K
Exterior Brickwork and concrete block masonry	1700Kg/m3	0.770Watts/metre K
Hardboard	1000Kg/m3	0.300Watts/metre K
Timber (hardwood)	700Kg/m3	0.180Watts/metre K
Timber (softwood), plywood, chipboard	500Kg/m3	0.130Watts/metre K
Soft fibre board	300Kg/m3	0.080Watts/metre K
Cork (varies in density and insulation value)	100Kg/m3	0.040Watts/metre K
Expanded polystyrene (EPS) board	15Kg/m3	0.040Watts/metre K
Mineral wool batt (eg Rockwool)	25Kg/m3	0.038Watts/metre K
Phenolic foam board (eg Kingspan)	30Kg/m3	0.025Watts/metre K
Polyurethane board (eg Roommate)	30Kg/m3	0.025Watts/metre K
Air (without convection)	1.2Kg/m3	0.023Watts/metre K

Note that the U values in the above table relate to a theoretical metre thickness for the purpose of calculation. In the hive, bees need to proactively control heat and ventilation to maintain the temperature they require. In harsh environmental conditions, improved insulation of the hive will allow them to do this at a reduced expenditure of their own energy and food reserves.

Comparison of a beehive made from examples of the above materials with modern building regulation for roof insulation:

25mm thick soft wood 0.130 / 0.020 = 6.5W/m2K

50mm thick expanded polystyrene 0.040 / 050 = 0.8W/m2K

50mm phenolic foam board 0.025 / 0.050 = 0.5W/m2K

Modern house roof to Building Regulations U value requirement = 0.18W/m2K min!

Selection of the most appropriate material and system

The balance of issues is not straightforward.

Hive materials

Straw

Not presently used in moveable frame hives, although it was in the nineteenth century, this is used in skeps and also the Sunhive. The insulation quality of straw is very close to that of polystyrene and is a natural product. The rough surface inside seems to be ideal for propolising, which helps the bees defences against disease, and being straw it can be built in many shapes and sizes. It is also very light. The majority of beekeepers who use straw for their bees, make their own skeps. On the downside, straw is not of itself waterproof without additional protection and cannot be sterilised. Straw is also attractive to mice and rats and straw hives need to be protected from them.

Wood

Hives are made of wood of different types and also plywood. The benefit of using wood is the beekeeper can make his own hive and adapt it to suit the circumstances. I am aware of home-made 12 frame deep Nationals, topbar hives made from recycled wood of varying lengths and so on. Wood is easy to sterilise by scorching. It is relatively easy to repair. Intended spacing can be imprecise both through human error and the use of unseasoned wood drying out or wood swelling. Hive walls can be made thick, as in the Layens hive, but the trade off is that it becomes very heavy.

It is also durable and robust allowing for years of use.

- **Cedar:** the lightest and most weather resistant. Favoured by beehive manufacturers across the UK.

- **Pine:** heavy and needs painting- dense and perhaps providing more insulation.

- **Plywood:** variable thicknesses and weights, marine ply is weather resistant. Plywood is made up of many fine layers of wood glued together into boarding. It is used in the Rose hive and sometimes the internal boxes of the WBC.

- **Recycled wood** is used, but caution as to prior treatments which may contain arsenic.

Polystyrene

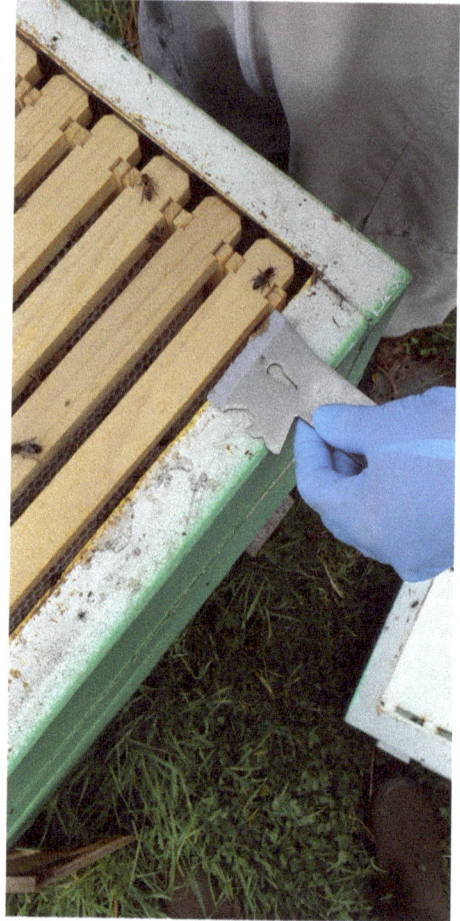

Left: a polystyrene hive; right: illustrating the thickness of the hive wall.

Polystyrene provides good insulation. It is apparent that colonies thrive in polystyrene and they are increasingly being used by professional beekeepers across Europe and Scandinavia. Hives made of polystyrene are moulded in factories, so cannot be adapted to individual uses as wooden hives. Some are made in sections which have to be glued together. Their spacing for beespace is fixed and will either be perfect or wrong.

Polystyrene is often lighter but always bulkier than wood. The denser the polystyrene the heavier the hive. Polystyrene nucleus hives are very popular with hobbyists for small colonies, and also for mating/ queen rearing nucleus hives . Because polystyrene, when empty can be light it is advisable to strap these hives to their stands to avoid the risk of wind damage. Alternatively some beekeepers use wooden lids on polystyrene hives.

Polystyrene needs to be painted to protect it from sunlight. Masonry paint is usually recommended. Polystyrene is not as robust as wood and can break. If maintained this material will last a long time.

Above / right: Waxmoth damage

Although it is easier to recycle polystyrene and uses less water than recycling paper, it is of note that its recycling remains an issue in the UK as there is presently no curbside collection for polystyrene, and not every council has recycling arrangements for it.

Cleaning polystyrene is also quite easy, but more time consuming than cleaning wood. Propolis has to be scrapped off with great care and the walls scrubbed with soda and then bleach. Soaking them in bleach in a wheelybin is a partial solution. However cleaning is not quick.

It is worth noting that at the present time no research has been done into the possible effects on honeybee health and populations when using polystyrene hives, but a study into the effects of polystyrene hives on honeybees is about to commence (2019) at Edinburgh Napier University.

Polypropylene (plastic)

There are two plastic hives on the market at the present time. One advantage of plastic over other materials is that wood peckers cannot make a hole in it.

The Beehaus (right), which is a plastic hive and similar to the Dartington Long Deep hive. This comes as a flatpack. It is insulated and requires careful assembly and posi-

tioning to avoid gaps. It uses standard 14 x
12 National frames

The Apimaye (right) is a commercial grade
plastic, heavily insulated Langstroth hive.
It is worth noting that it uses plastic Apimaye frames which are not the same size as
the standard available wooden Langstroth
frames in the UK, which would result in
major differences in beespace if used. It is
very heavy compared to a wooden equivalent, and has a large number of possible
add-ons and variations.

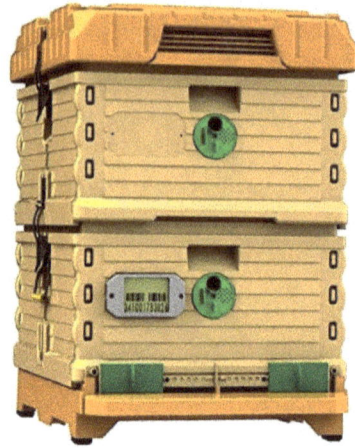

In common with polystyrene, plastic hives
need to be gently scraped cleaned and then scrubbed with soda and then soaked in strong
bleach in order to clean them adequately.

No research has been identified as to the effects on honeybee health and population of
using plastic.

Concrete breeze blocks

Insulated breeze blocks are cheap and easily available. The Zest hive is built with these, providing an insulated cavity for the colony. Breeze blocks are by definition heavy, and this hive
cannot be moved as a piece

Cork, a natural product

Not presently used to make hives but being marketed as a mean of insulating hive walls,
specifically for the Modified National and also the WBC at the present time.

Developments in Hive Components

2: Floors in response to pests

Defence against *Varroa*

Varroa mesh floor

With the arrival and naturalisation of *Varroa* destructor, thoughts turned to how beehive design might help in its management and monitoring. In about 2000 the open mesh floor was invented and has been promoted by the Bee Unit for many years. These have now become standard kit for many hives. They come complete with a tray which can be inserted to collect debris, in particular dead *Varroa* mites, and if the floor has been greased *Varroa* mite that have fallen off bees. This is used to monitor *Varroa* levels.

It is important to clean these boards off regularly or leave them out as the debris will attract ants, mice and wax moth.

They are very useful in the winter for identifying the size and position of the brood nest and whether or not the colony is eating stores without interfering with the bees directly.

Happy Keeper Floor

The tubes board was invented around 1993.

Since that time, experiments as well as beekeeper's observations have shown that a unique feature was at the origin of all advantages of this floor.

This feature is the possibility for the bees to control air renewal inside the hive, especially in winter, to make an atmosphere which is well suited to their needs. In summer, there are plenty of bees and they are able to drive used air outside the hive more or less easily through any opening. But it is not the same by cold weather when they are in cluster in the upper part. In that case, only the tubes floor is adequate.

A first observation can be made in February when there is already much brood although temperature is still cold. Condensation is then occurring on the tubes under the cluster showing that bees are expelling used air full of humidity under the hive.

A second observation can be made around the beginning of September, when natural falls of *Varroa* mites are starting, at least under colonies which have not been recently treated. At that time, if you change a mesh board with a tubes board, you can observe within a few minutes that mites are falling ten times more. And in that amount of time, only atmosphere inside the hive can have changed.

Mites are still falling during all winter which leads to very low infestation rates in spring. This must be controlled by new users with the method of washing bees, widely described on Internet. Under 5%, treatments are useless.

During the field trials near Vesoul (east of France) made with 20 colonies of black bees whose queens were sisters, it was also observed in March that the brood surface was practically double in colonies equipped with tube floors compared with colonies on mesh boards. In Toulouse, something equivalent was observed with colonies of Buckfast bees whose queens were also sisters: in May the brood surface was still 25% larger.

A considerable amount of work should have to be done to scientifically analyse what is precisely occurring inside the hive with different types of floor (solid floor, mesh floor, tubes floor) in terms of humidity rate, carbon dioxyde rate, temperature, air movements. But interesting results are already here! So that the tubes floor should be much more widely used for the health of the bees, to avoid the use of pesticides and for the productivity of the colonies.

Description written by Jean-Pierre LE PABIC who says: I have been keeping bees since 1992 and have been using the tubes boards since 1993. I have not applied any treatment of any kind since 1997. I have around 6 colonies which are in my garden in Rueil-Malmaison, 10 km far from Paris. I am using Dadant 12f hives.

Defence against hornets

In 2014 the Apishield hornet trap was invented in Greece, where it has been rigorously tested and used. This is a complex floor where the hornets and wasps are lured in through entrances under the hive into a self-contained box and are unable to escape.

Deterrent to mice and wasps

Thornes have recently produced a floor earlier very popular on the heather moors because it provides a sheltered entrance. It has been found to be a deterrent to wasps and mice.

Solid floor

Developments in Hive Components

3: Beeswax and foundation

Beeswax

All beeswax attracts dirt and absorbs fat-soluble chemicals as well as water soluble ones. Fatty chemicals are found in propolis and pollens, some pesticides, fungicides and some *Varroacides* including thymol. Foraging bees and treated bees can bring tiny amounts of these chemicals into the hive and they are absorbed into the wax comb. This keeps the colony safe from any ill effects. So, can this be a problem for the bees? It is only when bees chew the wax there is any risk of their coming into direct contact with the chemicals in the wax, so the likelihood is low. However, chemicals toxic to bees exist in the environment, even away from treated fields so all comb the bees build is likely to contain some level of contamination. The more the comb is used, the greater the contamination. The National Bee Unit strong advocate changing comb at least every three years primarily to reduce disease risk from the build of pathogens, but it also reduces the build-up of toxins in the hive.

Foundation wax

There have always been issues around the use of foundation. R.O.B. Manley expressed concern about the '*inevitable adulteration of the wax*'[1] in the 20th century. There is presently no means of removing the fat-based chemicals from recycled beeswax.

Foundation is always made from recycled wax. Beekeepers can make their own foundation from their own wax, but it a slow task and requires a lot of wax. It is a major saving of labour to the beekeeper to be able to exchange recycled beeswax for foundation made by a beekeeping supplier. It is made in many frame sizes, and embossed for worker, drone or small cell use. Thinner foundation is available for cut comb and most brood-frame sizes and sheets can be wired.

The beeswax used commercially to make foundation has always been collected from bee-keepers, either locally or from other countries. All beeswax is thoroughly cleaned by melting and allowing impurities to settle in a water bath. All water-soluble impurities are removed, just leaving the fatty chemicals. Commercially, some suppliers use acids to remove metal ions and some manufacturers may bleach the wax chemically. All wax that has been imported through British Wax has also been sterilised by heating to 120°C for a period of time. This would kill any fungal or disease spores that might be present.

So is it safe to use foundation? The short answer is yes. The chemicals trapped in the foundation are in microscopic quantities and are rapidly covered by the bees as they draw comb, and it seems sterilising is not needed. In New Zealand, where honey production is a core industry, their National AFB management plan investigated whether spores survived the process of foundation production and found no spores. They concluded that foundation is not an important factor in the spread of American Foul Brood disease.

However there remain unresolved issues of concern to beekeepers.

There are different chemicals prevalent in different countries, most worryingly Couma-phos, which is a *Varroacide* and can accumulate in wax. This is not used in the UK at present and is therefore not present in locally produced wax but might be in extremely tiny quantities in imported wax. Foundation normally described as 'standard' uses imported wax and foundation made from largely British and Irish wax is described as 'premium'.

In addition to this there is occasionally organic wax foundation. This is always imported and although collected many miles away from any agricultural activity may still contain traces of chemicals.

So why do beekeepers use foundation? It helps the bees is the short answer. Producing wax requires not only having the correct population mix, which a colony may not have, but also an enormous amount of energy which means a lot of nectar gathering, which is a weather and season dependent activity. It clearly also helps a beekeeper using a moveable frame hive. It encourages bees to draw comb in parallel so that frames used can be removed individually without harming the colony. Its use has also allowed bees to produce an excess of honey for the beekeepers to collect which facilitates honey production and was probably the trigger allowing for commercial honey production. Today many beekeepers minimise its use, alternating frames with foundation and frames with 'starter strips'. Some use none. The reasons for this is more to a perception that bees benefit from drawing their own comb rather than concerns about introducing toxins into the hive.

Wax embossing

Wax can be embossed in three cell sizes. Drone, worker and small cell. There have been several research projects pursuing the idea that small cells reduce *Varroa*. At present there has been no conclusive evidence that this is the case.

If a super frame is placed in a brood box between brood frames it is common for bees to draw drone comb below the bottom bar of the super frame.

Thin foundation

This is specifically for producing cut comb as standard foundation is quite thick.

Wired foundation

Before the advent of wired foundation frames had to be wired and the foundation if used had to be melted onto the wires. In larger hives this still remains the best way to ensure the comb remains stable in the frame when lifting frames out.

Plastic foundation

This is popular in the USA and is available for sale in the UK. Its advantage is that there is no need to buy replacement foundation because the comb can be cut off and the entire frame and plastic insert sterilised. This may be of particular benefit in supers for honey production on a large scale, although it is understood that at the present time it is not being adopted by commercial honey producers in this country. The other advantage is that it comes in either a yellow or black colour. The black if used in the brood nest makes seeing eggs easier in the brood nest.

There are however disadvantages. Firstly, each frame must be coated in wax before use.

This is a time-consuming operation, but probably only needs to be done once. The resulting foundation is thicker than beeswax foundation and is heavier.

Secondly there does not appear to have been any testing of the effects of these frames on honey bees. It is well recognised that honey bees use the comb surface to transmit communications, this is why they are thought not to attach the bottom edge of their comb to a surface, and why where foundation has been used, they often make holes around the base of the comb. It is of notes that some plastic foundation suppliers are now making sheets with sections that can been snapped off to allow for passage of bees.

References

- *Honey Farming* (page 72) – R.O.B. Manley, 1946
- *The Buzz about Bees* – J.Tautz, 2007

Developments in Hive Components

4: Hive monitoring

There was a time when the majority of our knowledge about honey bees came from watching the behaviour of bees at the hive entrance. Observation hives have allowed us to watch a colony through glass. Hive inspections allow us to exam a colony frame by frame. We have steadily intruded more and more into the life of the colony but still cannot always accurately interpret what we see! It is one thing to identify all stages of brood, sufficient stores and the presence of a queen, quite another to spot the colony is thinking about swarming, the colony is sick or the colony is under attack. Hive monitors place sensors inside a hive to measure all those things we cannot see through direct examination and send the information to a monitor. The way the measurements combine can be interpreted to explain what is happening, or has happened in the hive and this information can be accessed by the beekeeper.

Examples of events would be:

- Honey flow starting/ending
- Queen mating flights
- Swarm prediction/swarming
- Hive theft

It is thought that monitors will be soon be able to identify:

- Asian hornet attacks
- Nosema
- Brood monitoring

What can already be monitored?

- Weight
- Temperature
- Humidity
- Acoustics
- Weather

This idea has found great favour with research and academic institutions and is slowly gaining ground amongst hobby beekeepers. Not only might the beekeeper be able to 'keep and eye' on their bees without opening the hive, but information shared with researchers would allow for wide-scale information gathering and even greater understanding of bees. It is a developing technology and over the next few years is likely to become easier to use and more reliable. In the UK at present there are two providers: ARNIA and ApisTech. The information for this description has largely been provided by ARNIA.

Comparison of different hive designs

	No. of Frames	Brood Frame Size in inches	Maximum Number of Cells per Brood Box	Top/Bottom Beespace Warm/Cold	Lugs L/S
WBC	10	14 x 8 ½	52,000	B/W	L
WBC deep	10	14 x 12	73,450	B/W or C	L
Modified National	11	14 x 8 ½	57,100	B/ W or C	L
Deep National	11	14 x 12	78,550	B/W or C	L
Smith	11	14 x 8 ½	57,100	T/C	S
Rose	12	14 x 7		B/W or C	L
Dartington	22	14 x 12	2 x 78,550	B/C	L
Modified Commercial	11	16 x 10	74,250	B/W or C	S
Langstroth	8 or 10	17 ⅝ x 9 ⅛	72,250	T/C	S
Langstroth Jumbo	8 or 10	17 ⅝ x 11 ¼	93,700	T/C	S
Dadant	11	17 ⅝ x 11 ¼	93,700	T/C	S
Warré	8	Top bar approx. 12 inches		B/W or C	–
Topbar	Varies	Varies		Varies	–
Layens	14	17.3 x 13.38		B/C	S

Epilogue

Tricia Nelson

We have collected descriptions of just twenty-three hives in current use in the UK and Ireland, almost all designed by hobby beekeepers. Examples of materials in use, innovations in floor design, the changing use of beeswax and the development of monitoring have also been described. The history chapter identifies earlier hives which are no longer in use.

It is of note that with careful management and good forage the honey bees have survived in all these styles of hive! Perhaps this is why honey bee domicile has not been researched extensively until recently?

Hive design not only reflects our reasons for keeping bees but also our level of knowledge of honey bee needs, so it is not surprising that over the past few years there has been a further wave of ideas. There is little doubt that honey production has always been a reason to keep bees across the world, but there is no dispute their role in pollination is of far greater general concern to us all, and their existence has been a source of inspiration, wisdom and joy to many.

Honey bee pollination cannot be achieved without healthy colonies. The Foul broods were in part responsible for the interest in design in the 19th century, the effect of *Varroa* has perhaps been key in the present interest in alternative styles of beekeeping and with it hive design. However, all the other issues listed in the prologue still need to be considered together with the ever present threat of bacterial disease, namely American Foulbrood and European Foulbrood. There still remains much to discover about the optimum conditions for a hive in which honey bees can thrive. The comprehensive research of Professor Tom Seeley and other bee scientists into honey bee domicile has started the ball rolling, other researchers with different skill sets such as Derek Mitchell, a mechanical engineer, are also now researching this topic.

Yesterday I heard of another new design, too late for this publication, and tomorrow there may be another. The search for the perfect hive will no doubt go on. I hope you have enjoyed reading this book which is a snapshot of hives in use in the UK in 2019.

Acknowledgements

- Northern Bee Books

- Editorial team and DARG members for their assistance

- All the contributors

- Thornes

- British wax

- Donegal Bees

- Beekindhives for use of photographs

- Arnia

- Apimaye for use of photograph

- Omlet for use of photograph

- Thermosolarhive

- Flow™ Hive for use of photographs and illustrations